Jesus Finally Exposed

Experts Uncover Lies you thought were truth

Inspired by C. S. Lewis
Collaboratively written by Stephen Brown and Charles Gopatten
Using over 33 different experts, across the centuries,
to finally find the truth about Jesus

Contents

Many people skip the Introduction, so I made it one of the chapters (Chapter 0). This chapter gives you a good basis of what we are going to do.
You should read it.

Chapter 0

People's perception of a "Good" Person?

L et's play a game. I'll give you some quotes, and you try to match them to the right person. (No pressure, but your answers might surprise you.) Some names you'll recognize immediately. Some quotes? Maybe not so much.

Turn the page.

Match the Person to the Quote:
There may be more than 1 quote per person

1-Al Capone	A-"I am strongly in favor of using poisoned gas against uncivilized tribes."
2-Mahatma Gandhi	B-"Riots are the language of the unheard."
3-Winston Churchill	C-"Obstacles do not exist to be surrendered to, but only to be broken."
4-Pablo Escobar	D-"If someone rapes your sister, do not seek revenge. Offer him your other sister."
5-Adolf Hitler	E-"Be careful who you call your friends. I'd rather have four quarters than a hundred pennies."
6-Jesus	F-"Do not compare yourself to others. If you do so, you are insulting yourself."
7-Martin Luther King Jr.	G-"I will cause division between a man and his father, a daughter and her mother, and a daughter-in-law and her mother-in-law."
8-Niccolò Machiavelli	H-"Ideas are more powerful than guns."
	I-"Let the dead bury their own dead."
	J-"Life is full of surprises, some good, some not so good."
9-Joseph Stalin	K-"The ends justify the means."
	L-"The journey of a thousand miles begins with a single step."
10-Mao Zedong	M-"Whosoever desires constant success must change his conduct with the times."

Got your answers? Great. Check your answers.

Answer Key

1-Al Capone
E-"Be careful who you call your friends. I'd rather have four quarters than a hundred pennies."

2-Mahatma Gandhi
D-"If someone rapes your sister, do not seek revenge. Offer him your other sister."

3-Winston Churchill
A-"I am strongly in favor of using poisoned gas against uncivilized tribes."

4-Pablo Escobar
J-"Life is full of surprises, some good, some not so good."

5-Adolf Hitler
F-"Do not compare yourself to others. If you do so, you are insulting yourself."
C-"Obstacles do not exist to be surrendered to, but only to be broken."

6-Jesus
I-"Let the dead bury their own dead."
G-"I will cause division between a man and his father, a daughter and her mother, and a daughter-in-law and her mother-in-law."

7-Martin Luther King Jr.
B-"Riots are the language of the unheard."

8-Niccolò Machiavelli
K-"The ends justify the means."
M-"Whosoever desires constant success must change his conduct with the times."

9-Joseph Stalin
H-"Ideas are more powerful than guns."

10-Mao Zedong
L-"The journey of a thousand miles begins with a single step."

Now, let's talk about what just happened

So, Who's "Good" and Who's "Bad"?

You probably had some names pop out at you. Some were easy. Others? A little harder.

Now, be honest—did any of those quotes change your perception of the person who said them? (Or make you second-guess what you thought you knew?)

Take **Mahatma Gandhi**, for example. Most people think of him as a peaceful, moral leader. But if the only thing you knew about him was that **one** quote, what would you think? Maybe not so great.

That's how reputations work. A person can say or do many things, but we tend to judge them based on select pieces of information. And that's why, when it comes to forming opinions, we have to make sure we're using reliable sources— **truthful**, **accurate**, and **unbiased**.

Now, Let's Talk About Jesus

Most people say Jesus was a good man. Some even call Him a prophet.

But what if I told you He made claims that went far beyond anything Gandhi ever said? Claims that, if false, would put Him in an entirely different category?

If we're going to have an honest conversation about Jesus, we can't just rely on opinion. We need facts.

There are **ten major claims** about Jesus, and we're going to put them under the microscope. Not with preconceived ideas. Not with blind faith. Just **truth**—wherever that leads us.

Here's What We'll Explore:

- The Bible is not accurate
- Jesus was a liar
- Jesus was just a moral leader
- Jesus was just a legend
- Jesus was mistaken
- Jesus wasn't God
- Jesus was mentally ill
- Jesus didn't say what we think He said
- Jesus isn't Lord
- Jesus didn't rise from the dead

That's the list. Nothing's off-limits.

As an old TV detective used to say: **"Just the facts."**

Let's go.

Two more things.

We are going to be using the dating system BC and AD, rather than BCE and CE. BCE and CE (Before Common Era, Common Era) is not as well known.

All scripture is from the English Standard Version (ESV). The ESV is published by Crossway, a not-for-profit publishing ministry. We have placed the full scripture text in the back of the book for each chapter.

Chapter 1

The Bible Is Not Accurate

N ow, that's a bold statement. Maybe it made you sit up a little. For some, the Bible's accuracy is a challenge. For others, it's just a fact of life. But here's the big question: How do we know if *any* ancient story is true?

Historians and archaeologists have been at this for centuries— digging through manuscripts, unearthing artifacts, and basically playing the world's longest game of detective work. Their goal? To find out if the words we read today match what was originally written.

So, let's do a little investigation of our own. We'll start with some of the most famous ancient texts, see how they've held up over time, and then compare them to (you guessed it) the Bible. Let's go!

How Do Ancient Texts Hold Up?

Homer's *Iliad* and *Odyssey*

- **Date Written:** Around 750–700 BC

- **Oldest Copy We Have:** 300–200 BC

- **Time Gap Between Original and Copy:** 500 years

- **Copies Found:** Around 1,500

Even though the earliest *full* copies didn't show up until the medieval period, scholars compare fragments and references in other works to verify their accuracy. Long story short:

We're confident that what we read today is pretty close to what Homer actually wrote.

Tacitus' Annals

- **Date Written:** Around 110–120 AD
- **Oldest Copy We Have:** 900 AD
- **Time Gap:** 800 years
- **Copies Found:** 2 major manuscripts + fragments

Despite missing sections, Tacitus remains a go-to source for Roman history. His work survived mostly thanks to monastic scribes who apparently didn't mind his criticism of emperors (brave monks).

Herodotus' Histories

- **Date Written:** Around 430–420 BC
- **Oldest Copy We Have:** 900 AD
- **Time Gap:** 1,300 years
- **Copies Found:** About 100

Herodotus, often called the "Father of History," had a reputation for mixing facts with the occasional tall tale. But archaeology has backed up a surprising number of his claims—so even with embellishments, his work holds historical value.

Shakespeare's Hamlet

- **Date Written:** Around 1600 AD
- **Oldest Copy We Have:** 1603 AD (First Quarto)
- **Time Gap:** 3 years

- **Copies Found:** About 235 early editions

Since *Hamlet* was printed soon after it was written, scholars are confident we're reading what Shakespeare actually intended (tragic monologues and all).

What About the New Testament which speaks of Jesus in the Bible?

The New Testament

- **Date Written:** 50–100 AD
- **Oldest Copy We Have:** 125 AD
- **Time Gap:** 25–75 years
- **Copies Found:** About 5,800 in Greek (plus thousands more in other languages)

The New Testament far surpasses every other ancient text in terms of preservation, accuracy, and sheer volume of manuscripts. The earliest New Testament manuscripts come from within a few decades of the original writings, an extraordinary feat compared to other historical texts. Because of this, scholars have been able to reconstruct the New Testament with over 99% accuracy.

The New Testament contains accounts of Jesus' life, teachings, and resurrection, with multiple independent sources confirming the same events. The level of consistency between manuscripts and translations over centuries solidifies its credibility as a historical document.

What Does Archaeology Say?

History is great, but let's get physical. Artifacts, ruins, and inscriptions help us determine whether ancient writings were based on real places, real people, and real events.

Homer's Iliad and Odyssey

Some believed these stories were purely myth. But then archaeologists discovered the ruins of Troy, proving that at least some of Homer's tale was based on real events.

Tacitus' Annals

Archaeologists have found burned city layers, Roman coins, and inscriptions that confirm Tacitus' accounts of Rome. His writings weren't just opinion pieces—he recorded actual history.

Herodotus' Histories

Skeptics doubted Herodotus, but then they found the Behistun Inscription, an ancient rock carving that matched his accounts of Persian conquests. Then there's the Ruins of Babylon—right where he said they'd be.

The Bible's Archaeological Backing

The Bible has been continuously supported by archaeological finds. Here are just a few examples:

- **The Dead Sea Scrolls (250 BC – 68 AD):** Confirmed the Old Testament had been faithfully copied for over a millennium.

- **The Walls of Jericho (1400 BC):** Excavations reveal a sudden collapse, just as described in Joshua 6.

- **The Caiaphas Ossuary (1st Century AD):** A bone box confirming the high priest who tried Jesus was real.

- **The Pilate Stone (1st Century AD):** Confirms that Pontius Pilate, who sentenced Jesus to death, was the Roman governor of Judea.

- **The Tel Dan Stele (9th Century BC):** A stone inscription mentioning the "House of David," proving King David wasn't just a legend.

- **The Pool of Siloam (Discovered in 2004, Jerusalem):** Proves John 9 describes a real place.

- **The Hezekiah Tunnel (701 BC, Jerusalem):** Contains an inscription confirming biblical accounts of Hezekiah's preparations against the Assyrians.

- **The Babylonian Siege of Jerusalem (587 BC):** Burn layers, clay tablets, and Babylonian records confirm this biblical event.

- **The Nazareth Inscription (1st Century AD):** A Roman decree banning grave robbing—possibly linked to early Christian claims of Jesus' resurrection.

So, Is the Bible Inaccurate?

If we accept the *Iliad*, *Annals*, and *Histories* as reliable (despite their gaps)... If we accept that Shakespeare wrote his plays, though only 250 original copies have been found... then dismissing the Bible doesn't make much sense. The Bible—especially the New Testament—has more copies, shorter time gaps, and stronger archaeological backing than any other ancient text.

At the end of the day, you get to decide what you believe. But if you're looking at the evidence, it's pretty hard to call the Bible "inaccurate." Just sayin.

Chapter 2
Jesus Was a Liar

Alright, let's talk about something a little controversial. Was Jesus a liar? Some people think so. Maybe he made stuff up. Maybe he exaggerated. Maybe he was playing some kind of long game. And if any of that were true—if Jesus intentionally deceived people—well? Let's dig in and see where the evidence leads.

The Premise of the Argument

The argument goes like this: Jesus said he was the Son of God, the Messiah, the Savior of the world. If that wasn't true, then he was either lying or completely delusional. Some skeptics argue that he made false claims to gain influence, misled his followers, or manipulated situations to serve his agenda. But here's the catch—if we examine his words and actions closely, the idea of Jesus as a liar starts to fall apart.

Evidence of Deception (or Not?)

There are a few passages that critics love to point to when making their case. Let's break them down.

- **John 7:8-10** – Jesus tells his brothers he isn't going to the Feast of Tabernacles, then later shows up. Sounds sneaky, right? But hold on. He didn't say, "I will never go." In some manuscripts he said, "I am not going up *yet*". When he does go, he goes in secret, not in the way his brothers expected. Not a lie—just a different approach.

- **Mark 4:10-12** – Jesus says he speaks in parables so people will "hear but not understand." If you take this at face value, it seems like he's deliberately hiding the truth. But dig a little deeper, and you'll see that Jesus is challenging people to seek truth, not just spoon-feeding them information. He's not misleading; he's filtering for those who truly want to understand.

- **Matthew 24:34** – "This generation will certainly not pass away until all these things have happened." Some argue that this was a false prophecy. But "generation" can mean a lot of things in biblical language—like a group of people or an era. And the destruction of the temple in 70 A.D.? That happened within a generation. Again, not a lie—just a different way of speaking.

Evidence of Truthfulness

Now, let's flip the script. If Jesus was a liar, you'd expect to find moments where he contradicts himself or behaves in a way that suggests deception. But instead, we get this:

- **John 14:6** – "I am the way, the truth, and the life." That's a bold statement. You don't make that kind of claim unless you truly believe it (or you're just reckless).

- **John 18:37** – Jesus stands before Pilate and says, "For this purpose I was born and for this purpose I have come into the world—to bear witness to the truth." If he was lying about his identity, this would've been the perfect moment to backpedal and save his skin. But he doesn't.

- **Matthew 5:37** – "Let what you say be simply 'Yes' or 'No'; anything more than this comes from evil." If Jesus was all about truthfulness, why would he contradict his own teachings? That wouldn't make sense.

Why Would Jesus Lie?

If we even entertain the idea that Jesus was lying, we have to ask—why? What was his motivation?

- **To protect himself?** Nope. If he wanted to avoid trouble, he wouldn't have kept saying things that got him into trouble (see: the whole crucifixion thing).

- **To gain power or wealth?** Not a chance. He rejected political power, had no interest in money, and constantly taught about humility and service.

- **To avoid embarrassment?** If that were true, he wouldn't have openly associated with sinners and social outcasts.

- **To protect others?** Maybe, but if that were the case, he wouldn't have warned his followers that they'd face persecution for believing in him.

Historical Context

Some people argue that maybe Jesus wasn't a liar, but he exaggerated or spoke in ways that his audience misunderstood. And that's fair—first-century Jewish culture used a lot of hyperbole and indirect communication. But even non-Christian historical sources, like Josephus and Tacitus, describe Jesus as a moral teacher, not a fraud.

- **Josephus** – A Jewish historian who described Jesus as a wise teacher who performed "surprising deeds."

- **Tacitus** – A Roman historian who mentioned Jesus' execution under Pontius Pilate, but never called him a deceiver.

Even those who opposed Jesus never really accused him of lying. They called him a blasphemer, a troublemaker, or a threat—but not a liar.

Let's be real

If Jesus was a liar, how could he be moral? The New Testament presents him as the embodiment of truth. His resurrection is meant to be the ultimate proof that he was who he said he was. If he made it all up, his teachings lose their credibility, and billions of people have been following a con artist for over 2,000 years.

But let's say he wasn't lying. Let's say he actually was who he claimed to be. In that case, we're not just dealing with a good teacher or a wise man. We're dealing with someone who changed history forever.

So, was Jesus a liar?

The evidence just doesn't support it. Sure, some passages might seem tricky at first glance, but a closer look shows consistency in his words and actions. He didn't seek power, wealth, or self-preservation. He didn't waver under pressure. His life, teachings, and historical impact point to someone deeply committed to truth, even when it cost him everything.

In the end, you don't have to believe that Jesus was the Son of God to see that he wasn't a liar. His message, his character, and even his enemies suggest that he was, at the very least, an honest man who lived - and died - for what he believed in.

Chapter 3

Jesus Was Just a Moral Leader

L et's be honest—Jesus is a big deal. Whether you believe he was divine or just a really wise guy, you can't ignore his influence. Some say he was just a great moral teacher—kind of like a first-century version of a motivational speaker with a really strong following. Others say he was much more than that. Let's move forward..

Ethical Teachings

Jesus had some pretty incredible moral insights. Love your enemies? Turn the other cheek? The Sermon on the Mount (Matthew 5-7) was basically the greatest TED Talk ever, packed with wisdom on humility, mercy, and treating others with kindness. The Golden Rule—"Do to others as you would have them do to you." (Luke 6:31)—isn't just a nice saying; it's foundational for ethical living across cultures.

But here's the kicker—Jesus didn't just talk about being good. He spoke with authority. He didn't quote others to back himself up. Instead, he said things like, "You've heard it said... but I say to you" (Matthew 5:21-48). That's bold. And then there were the miracles—healing the sick, calming storms, even bringing people back from the dead. Not your average moral teacher.

Social Justice

Jesus wasn't just about wisdom; he was about action. He hung out with outcasts—tax collectors, women, the poor, and the sick. He flipped tables when people turned the temple into a marketplace (Matthew 21:12-13). He challenged the religious elite. This wasn't a guy who stayed in safe spaces—he went where people were hurting and lifted them up.

But here's the difference: Jesus wasn't just advocating for social justice in a general sense. He claimed he was saving humanity. " For the Son of Man came to seek and to save the lost." (Luke 19:10). That's a much bigger mission than just improving society—he was claiming divine authority to change eternity.

Personal Virtues

Let's talk about Jesus' character. Compassion? Off the charts. Humility? He washed his disciples' feet (John 13:1-17), which, let's be honest, was probably not a pleasant experience in the days of dirt roads and open-toe sandals. Forgiveness? He forgave the very people who nailed him to a cross (Luke 23:34).

But here's where things go beyond just "good teacher" territory—he also forgave sins. Not just people who had wronged him personally, but sins in general (Mark 2:5-7). That's not something a normal moral leader does. He claimed oneness with God (John 10:30) and accepted worship (Matthew 14:33). Either he was telling the truth, or he was saying things that no regular wise man should say.

Impact on Society

You don't have to be religious to acknowledge that Jesus' influence is staggering. His teachings have shaped entire civilizations. Laws, ethics, human rights movements—so many trace their roots back to his words.

But again, it's not just about moral influence. The doctrines of the Trinity and Incarnation—God becoming human—set Christianity apart. Jesus wasn't just a philosopher or a social reformer. His followers believed he was God in the flesh. That changes everything.

Comparison with Other Moral Figures

Stack Jesus up against other moral teachers like Confucius, Buddha, or Socrates, and you'll see similarities—wise sayings, ethical principles, impactful followers. But there's one glaring difference—none of them claimed to be God. None of them predicted their own death and resurrection—and then pulled it off.

Also, Jesus' followers didn't just admire his ideas; they literally died believing he was the Son of God. Many were tortured, executed, or exiled. You don't get that level of commitment for just a wise teacher. People might die for an ideology, but they don't usually die for something they know is a lie.

So, was Jesus just a moral leader?

If you've stuck with me this far, you probably see where this is going. Yes, he was a brilliant teacher. Yes, he modeled extraordinary kindness, justice, and virtue. But he also made claims about himself that no mere moral leader would make. His followers didn't just preserve his teachings; they gave their lives proclaiming that he was the Son of God. And if he really did rise from the dead, then calling him "just a moral leader" is a pretty big understatement.

Jesus Finally Exposed

Chapter 4
Jesus was a Legend

S
ome say Jesus was nothing more than a legend—his story woven and re-woven until it became something larger than life. Others insist he walked, spoke, taught, and changed history in ways too significant to dismiss. This chapter sifts through the arguments, examining historical methods, religious comparisons, textual scrutiny, and cultural influences. The goal? A conclusion rooted in evidence.

How Historians Work

Determining historical reliability isn't guesswork. Scholars apply the historical-critical method, dissecting authorship, timeframes, and cross-referencing sources. The goal? Separating fact from interpretation.

Three non-Christian sources—Tacitus, Suetonius, and Josephus—mention Jesus. Each comes from a different background, yet all contribute to the discussion of Jesus as a historical figure.

Tacitus, a Roman senator and historian, wrote *Annals* around 116 AD. His reputation for accuracy and skepticism makes his work particularly valuable. When he describes Christ's execution under Pontius Pilate and Emperor Nero's persecution of Christians, he does so not as a supporter but as a critical observer. Tacitus had access to Roman state records, reinforcing the reliability of his statement.

Suetonius, another Roman historian, was a scholar and biographer under Emperor Hadrian. In *The Twelve Caesars* (c. 121 AD), he notes disturbances in Rome linked to "Christus," a term often linked to Christ. Though his reference is brief, it signals Jesus' influence extending beyond Judea and into the political world of Rome within decades of his death.

Josephus, a Jewish historian born in 37 AD, provides two references to Jesus in *Antiquities of the Jews* (c. 93-94 AD). His background is significant—he wasn't a Christian but a Pharisaic Jew with Roman ties. His first reference, the *Testimonium Flavianum*, describes Jesus as a wise teacher crucified under Pilate. Some wording suggests later Christian edits, but a more neutral core text likely remains authentic. His second mention is more direct: he identifies James as "the brother of Jesus, who was called Christ." That passage is widely accepted as genuine.

Beyond these three, additional sources hint at Jesus' presence in history. **Pliny the Younger**, a Roman governor, wrote to Emperor Trajan in 112 AD about Christians worshiping Christ as divine. The *Babylonian Talmud*, a Jewish text, references the execution of "Yeshu," aligning with Jesus' known timeline.

These sources, independent of Christian tradition, bolster the argument for a historical Jesus. They suggest that, whether divine or not, Jesus lived, taught, and was executed—a conclusion built not on faith, but on historical evidence.

The Comparative Lens

Legends have a way of forming around historical figures. Many religious traditions feature themes of sacrifice, resurrection, and salvation. Figures like Mithras, Osiris, and Krishna share thematic similarities with Jesus. The temptation? To lump them all together, to reduce Jesus' story to another iteration of ancient mythology.

But history has a way of distinguishing between myth and fact.

Embarrassment

Multiple independent sources reference Jesus' crucifixion, reinforcing its authenticity. The criterion of embarrassment—the idea that followers wouldn't invent something unfavorable—supports it further. A crucified Messiah? That's not the hero they would've scripted. In a world where power was revered, the idea of a suffering, executed savior was counterintuitive. The early Christians had every reason to downplay it—but they didn't.

"The God of Abraham, the God of Isaac, and the God of Jacob, the God of our fathers, glorified his servant Jesus, whom you delivered over and denied in the presence of Pilate… you killed the Author of life, whom God raised from the dead. To this we are witnesses." (Acts 3:13–15)

Multiple Attestation

Then there's the principle of multiple attestation. The Gospels record Jesus' death, but so do Paul's epistles, non-Christian Roman sources, and Jewish historical writings. The crucifixion is not a lone account but a repeated historical claim from different, often opposing perspectives.

"But the angel said to the women, "Do not be afraid, for I know that you seek Jesus who was crucified. He is not here, for he has risen, as he said. Come, see the place where he lay. Then go quickly and tell his disciples that he has risen from the dead, and behold, he is going before you to Galilee; there you will see him. See, I have told you.""(Matthew 28:5–7)

His teachings also stand apart. Many mythological figures are shaped by the values of their time, but Jesus' messages often ran against the grain. *Love your enemies. Bless those who curse you. The first shall be last.* These teachings didn't fit the mold of a conquering Messiah or the expectations of Rome. They were radical. Disruptive. And they caught on because they were different.

"But I say to you who hear, Love your enemies, do good to those who hate you, bless those who curse you, pray for those who abuse you."(Luke 6:27–28)

Mythology

Comparisons to other mythological figures fall short when examined under historical scrutiny. Mithras, for example, was a god of the Roman mystery cults, worshiped primarily by soldiers. His supposed resurrection? The evidence is weak. Osiris? His "resurrection" was tied to the agricultural cycle, symbolizing renewal rather than an actual historical event. Krishna? His divine exploits belong to Hindu mythology, not recorded history.

Jesus' story unfolds in a real historical setting, among real historical figures, in a place and time verified by external sources. The comparisons to mythical gods fail to account for this. Jump back to Homer, they thought Troy was a myth until they found it.

So, was Jesus merely a legend, stitched together from past traditions? Or does the weight of history point to something more?

"So Pilate, wishing to satisfy the crowd, released for them Barabbas, and having scourged Jesus, he delivered him to be crucified."(Mark 15:15)

Just more of the same

Critics argue that Jesus' story borrows from earlier myths. Some say he was a composite—a blend of messianic figures. Others believe Christianity spread for sociopolitical reasons, not because of a historical Jesus.

Yet history resists that conclusion. Archaeology uncovers names, places, and figures that align with Gospel accounts. The **Pilate Stone** confirms Pontius Pilate's existence. The **Nazareth Inscription** suggests early Christian beliefs. The spread of Christianity wasn't just a matter of political convenience—it was a movement propelled by people who insisted they had encountered something real. A man. A teacher. A crucifixion. And what they claimed happened after.

"For we did not follow cleverly devised myths when we made known to you the power and coming of our Lord Jesus Christ, but we were eyewitnesses of his majesty."(2 Peter 1:16)

The disciples had every reason to abandon the story. Persecution. Imprisonment. Death. But they didn't. Instead, they traveled, preached, and spread a message they believed was worth everything. Legends don't tend to birth movements that endure oppression. Something happened that convinced them Jesus wasn't just another name lost to history.

"and when they had called in the apostles, they beat them and charged them not to speak in the name of Jesus, and let them go. Then they left the presence of the council, rejoicing that they were counted worthy to suffer dishonor for the name. And every day, in the temple and from house to house, they did not cease teaching and preaching that the Christ is Jesus."(Acts 5:40–42)

Legend or Not?

Jesus wasn't just a legend. He was a real figure in history. His story, wrapped in theological meaning, still stands on the foundation of historical evidence. His name appears not just in sacred texts, but in the annals of those who had no reason to fabricate his existence.

Was he divine? That's a different debate. But the evidence that he lived, taught, and was crucified under Pilate is strong. The weight of history affirms his presence. The debate may continue, but the conclusion remains: **Jesus existed. And history remembers.**

Chapter 5

Was Jesus Just Mistaken?

L et's face it...There are some big questions in life. Like, why does coffee taste better on rainy mornings? Or why do we always pick the slowest checkout line at the store? But today, let's tackle a question that's even bigger (way bigger): Was Jesus mistaken about who he was, or did he know exactly what he was saying?

This isn't just some interesting thought experiment. It's a huge deal. If Jesus was just a good guy who got a little carried away with his claims, that's one thing. But if he truly knew he was divine—if he wasn't just hopeful or self-deceived—then that changes everything. So, let's take a step back, grab our metaphorical detective hats, and look at the evidence from every angle.

The Psychological Side of Things

Jesus as Sincerely Mistaken

Okay, so what if Jesus was just... wrong? It happens. Even the best of us can misinterpret things (GPS directions that send us into a lake). Some scholars suggest that Jesus may have genuinely believed he was divine but was influenced by psychological and social factors. He lived in a world soaked in Messianic hope, and people around him wanted to believe he was the real deal. Maybe he convinced himself of it, too. We've seen charismatic leaders throughout history do the same thing.

Jesus as Knowing and Truthful

But there's another side to this. If Jesus were just mistaken, wouldn't there be some cracks in his story? Wouldn't his teachings feel, well... off? Instead, we see a level of wisdom, insight, and moral teaching that still holds up today. He wasn't some wild, delusional prophet yelling on a street corner. He spoke with calm authority, challenged the smartest scholars of his day, and even predicted his own death and resurrection. That's not the profile of someone who's just fooling himself.

Cultural Influences—Did Jesus Buy Into the Hype?

The Messianic Expectations

Jesus wasn't the only guy people thought might be the Messiah. There were others before and after him who got swept up in the excitement, only to fade into obscurity. So, was Jesus just another hopeful leader who misread the signs?

Not so fast. Unlike others, Jesus didn't just play into the existing expectations. He flipped them. People wanted a warrior king to overthrow Rome. Jesus preached about loving your enemies and turning the other cheek. That's not how you win a political revolution, but it is how you change hearts. If he was mistaken, he sure was consistent in being a very *different* kind of mistaken.

The Consistency and Coherence of His Teachings

If you've ever tried to keep a lie straight, you know how hard it is. One slip-up, and the whole thing unravels. But Jesus' teachings hold up—across different accounts, different writers, and different situations. Whether he was talking to fishermen or scholars, his message remained the same. If he were simply confused about his identity, we'd expect contradictions or moments of uncertainty. Instead, his words and actions stayed remarkably aligned.

Scriptural Analysis—What Do The Texts Say?

Moments of Doubt

Now, to be fair, there are moments in scripture where Jesus seems to express uncertainty—like in Gethsemane, where he prays, "Remove this cup from me" (Luke 22:42). But is that a sign of doubt about his identity or just the natural human struggle of someone facing unimaginable suffering?

It's one thing to doubt who you are; it's another thing entirely to know exactly who you are but wrestle with the cost of it. That night in the garden wasn't about Jesus second-guessing his mission—it was about feeling the full weight of what was coming. He knew what lay ahead: betrayal, torture, a brutal death. And yet, even in that moment of agony, he followed it up with, "Yet not my will, but yours be done." That's not someone unsure of himself—that's someone fully aware, fully committed, and fully human in his suffering. If anything, his prayer in Gethsemane reinforces his identity rather than weakens it. Because real strength isn't pretending suffering doesn't exist—it's walking straight into it with eyes wide open. If Jesus was having an identity crisis in Gethsemane, he had a funny way of showing it—because instead of running, bargaining, or backing down, he stared suffering in the face and said, 'Let's do this.'

Foreknowledge and Authority

Then there are the passages where Jesus flat-out claims divinity. He calls himself the "I AM" (John 8:58), forgives sins (Mark 2:5-7), and tells people that seeing him is like seeing God (John 14:9). If he was mistaken, he was incredibly bold about it. And here's the kicker—his followers believed him, not just for a moment, but for the rest of their lives. Even under persecution. Even facing death. That's not how people act if they suspect their leader got it wrong.

Why It Matters

What If Jesus Knew Exactly Who He Was?

If Jesus knew exactly what he was saying—if he wasn't mistaken—then that changes the game. It means his words have divine weight. It means his promises are trustworthy. It means that the cross and the resurrection weren't just tragic misunderstandings, but part of a plan designed from the start. And that means we're dealing with a truth that's worth everything. If Jesus knew exactly who he was, then every word he spoke wasn't just hopeful encouragement—it was the kind of truth that flips tables, changes history, and demands a response.

What Do People Say?

The Skeptical View

Some skeptics say Jesus' followers exaggerated his claims after his death. Maybe he was just a great teacher, and over time, people made him into something bigger than he really was. It's not a bad theory. But it runs into a huge problem— why would his closest followers, who knew him best, be willing to suffer and die for something they knew wasn't true? If the disciples had been making stuff up, you'd think at least one of them—when faced with prison, torture, or execution— would've said, 'You know what? Never mind, we were just exaggerating,' but instead, they doubled down, which says a whole lot.

The Power of Testimony

History is filled with people who die for what they *believe* to be true. But people don't willingly die for something they *know* is a lie. The early disciples weren't just spreading a mistaken identity—they were giving their lives for something they had *seen* firsthand. That's a different level of conviction. If Jesus had been mistaken, surely someone along the way would have cracked under pressure. Instead, they stood firm.

If Jesus was mistaken, you'd expect at least one of those early followers to cash out and walk away—but instead, they ran straight into the fire, which tells you something real was at work.

The Final Question: Was Jesus Mistaken or Did He Know?

So, here's where we land. If Jesus was just mistaken, we'd expect to see inconsistencies in his teachings, doubts in his followers, and a gradual fading of his influence over time. Instead, we see a message that holds together, a group of believers who refused to back down, and a movement that has shaped the world for over 2,000 years.

Could he have been mistaken? Sure. But the evidence points in another direction. A direction that says he knew exactly who he was, exactly what he was saying, and exactly why it mattered. And if that's the case... then everything he said is worth taking seriously.

Jesus Finally Exposed

Chapter 6

Jesus wasn't God

So, Jesus—God or not? That's the question, right? It's one that has fueled centuries of debate, divided theologians, and probably made for some tense family dinners. And if you're here, you've probably wondered about it too (maybe more than once).

This isn't just some abstract theological discussion for people in robes sitting in stone buildings (though, yes, they've weighed in too). This is foundational—like, if Jesus *is* God, that changes everything about how we see Him, follow Him, and, frankly, build our entire lives. If He *isn't*, well... that's a game-changer too.

So, let's take a fair look at both sides. No yelling, no eye-rolling, no theological food fights—just an honest exploration of what Scripture, history, and logic tell us.

The Unitarian Take: Good Guy, Great Teacher, But Divine? Not So Much.

Some folks—Unitarians, many Jewish and Muslim scholars, and even some early followers of Jesus—argue that while Jesus was an incredible teacher, He never claimed to be divine. In fact, they say He pointed people to God, not to Himself.

Here's the big one: Nowhere in the Gospels does Jesus explicitly say, "Hey everyone, I am God. Worship me." (And yeah, that *would* have cleared some things up.) Instead, He talks about *the Father*—praying to Him, obeying Him, pointing others toward Him.

And that famous Sermon on the Mount? Pure gold—but it's all about loving God, not about Jesus claiming divinity.

Then there's history. Many scholars argue that Jesus' first followers—people who walked, talked, and ate with Him—didn't see Him as God, just as the Jewish Messiah. They say the whole "Jesus is divine" thing developed over time, shaped by cultural influences (hello, Greco-Roman world) and church councils.

It's a compelling argument. And, honestly, if you stop there, it makes some sense. But... (and you knew there was a "but" coming) ...there's another side to this.

Did Jesus Really Never Claim Divinity? Well...

While Jesus may not have spelled it out in flashing neon letters, He did drop some major hints—ones that got Him in *a lot* of trouble.

Take John 8:58. Jesus says, *"Before Abraham was, I am."* That's not just flowery language—that's a direct reference to God's name in Exodus 3:14 (*"I AM WHO I AM"*). The people listening got the message loud and clear—they tried to stone Him on the spot for blasphemy.

And what about Philippians 2:6-7? Paul, writing only a few decades after Jesus' death, says that Jesus was "in very nature God" but chose to humble Himself and take on human form. That's not some later church-invented doctrine—this was the belief of the earliest Christians.

Cultural Influences or Something More?

Some argue that Jesus' divinity was influenced by Greek and Roman thinking—you know, how they turned their emperors into gods. But here's the thing: the first Christians? They were Jewish—people who were *allergic* to anything resembling polytheism or idolatry. The idea that they'd suddenly decide to deify a man goes against everything they believed.

And yet... they did. Not because they were pressured by Rome (Rome was, in fact, killing them for it), but because they were convinced Jesus was *exactly who He said He was.*

The Early Church: Sorting It All Out (Or, At Least Trying To)

The early Christian church had some *major* discussions about Jesus' divine nature. (And by "major," I mean some of these debates involved exile, political drama, and, you know, the occasional excommunication.)

Fast forward a few centuries, and look at the biggest controversy. Arius—a theologian from Alexandria—argued that Jesus wasn't co-eternal with God but was instead *created* by Him. In other words, Jesus was important, sure, but not on the same level as God. It was called the Arian debate in the 4th century.

The Church's Role: Did They Just Make This Up in Turkey?

In Nicaea, Bithynia (currently İznik, Turkey) the church convenes the Council of Nicaea (325 AD) officially declares Jesus is divine. Some say that's when the council *decided* Jesus was God, but history tells a different story.

Church leaders weren't inventing new theology. They carefully reviewed scripture, including words of the disciples. And long before Nicaea, church leaders were already saying Jesus was divine. **Ignatius of Antioch (110 AD)** referred to Jesus as "our God" in his letters (which, let's be honest, seems like a pretty strong statement). **Justin Martyr (150 AD)** defended Jesus' preexistence. And **Irenaeus of Lyons (180**

AD) was all-in on the idea that Jesus was both fully God and fully man.

After the council the Nicene Creed was adopted. Stating that for Christianity, Jesus was God **("Begotten, not made, being of one substance (*homoousios*) with the Father.")** Also, after the council Arianism faded significantly, because believers could see the truth and logic that was presented to them,

So, nope - Jesus' divinity, him being God, wasn't a "political decision" cooked up in the 4th century. The council of Nicaea didn't invent it. They were preserving what had already been believed—long before Rome got involved.

How about some Tricky Theological Questions

Alright, so if Jesus *is* God, some tough questions come up:

- **Why did He pray?** If He's God, why talk to Himself?

    o Answer: Jesus, in His human form, modeled what it meant to have a relationship with God. He wasn't talking to Himself—He was showing us how to depend on the Father.

- **Why didn't He know everything?** (Mark 13:32—" But concerning that day or that hour, no one knows, not even the angels in heaven, nor the Son, but only the Father.")

    o Answer: Philippians 2:7 says Jesus voluntarily limited Himself. He was still God, but He laid down some divine privileges to walk fully as a human.

- **How can He be 100% God and 100% human?** That math doesn't check out.

    o Answer: That's the mystery of the **Hypostatic Union**—fully God, fully man, no blending, no switching between the two. It's a divine paradox, but not necessarily a contradiction.

But Wait—Wasn't the Messiah Supposed to Be Just a Leader?

Alright, let's switch gears. Some folks argue that Jesus, as the Messiah, wasn't supposed to be divine. Just a great leader. A chosen one. The Jewish understanding of the Messiah didn't necessarily include "God in the flesh" as part of the job description.

And that's… partly true.

But (and it's a big but) some Jewish scriptures and traditions hint at something *more*—a Messiah with divine qualities. Consider these:

- **Daniel 7:13-14** talks about "one like a Son of Man" coming with the clouds and having *everlasting dominion*. That's not just an ordinary human king—that's something far greater. Jesus actually claimed this title for Himself. (Not exactly a low-key move.)

- **Isaiah 9:6** calls the coming Messiah "Mighty God" (El Gibbor) and "Everlasting Father." Well, that's… kind of a big deal.

- **Psalm 110:1** places the Messiah at God's right hand—another serious position of divine authority. Jesus referenced this passage to make a point about His own identity.

- **Intertestamental Jewish Writings (like 1 Enoch, 4 Ezra)** describe a heavenly, preexistent Messiah figure. Meaning some Jewish groups were already thinking of the Messiah as more than just an earthly leader.

So, while *some* Jewish expectations saw the Messiah as purely human, others clearly anticipated someone *much more*. And Jesus? He fit those bigger, more divine descriptions *exactly*. By the way, Jesus did claim to be the Messiah. John 4:25-26 The woman said to him, "I know that Messiah is coming (he who is called Christ). When he comes, he will tell us all things." Jesus said to her, "I who speak to you am he."

The Bottom Line

Was Jesus divine? Early Christians believed so—passionately. And not just because some church council told them to. They believed it because:

- **Jesus Himself made claims that pointed to His divinity** (John 8:58, anyone?).

- **His earliest followers worshiped Him as God** (and let's be real—devout Jews didn't just *do* that without a serious reason).

- **Early church leaders taught it long before councils and creeds made it "official."**

- **Even some Jewish traditions left room for a divine Messiah.**

At the end of the day, the belief in Jesus as God wasn't a late-breaking theological twist. It was there from the start—bold, dangerous, and life-changing.

And honestly? It still is.

So, Where Does That Leave Us?

Jesus' divinity isn't just a theological question—it's *the* question. If He really is God, then everything He said about salvation, eternity, and our relationship with Him is solid, unshakable, and worth building our entire lives on. But if He *isn't*? Well, then Christianity collapses under its own weight.

There's no middle ground. Either He is who He claimed to be, or He isn't. And if He *is*, then that changes everything.

So, let's get into it.

Jesus Is God (And the Bible's Pretty Clear About It)

If you want to know whether Jesus is God, the best place to start is **what He actually said about Himself**. And spoiler alert: He didn't exactly leave it up for debate.

Scriptural Evidence (Because, Yes, It Matters)

The New Testament *over and over* points to Jesus' divinity. Not in some vague, symbolic, "Jesus is a great guy" kind of way, but in clear, bold statements. Let's look at some of the strongest ones.

John 8:58 – "Before Abraham Was, I AM"

- Jesus didn't just drop theological truth—He dropped *mic-level* truth.

- He said, "Before Abraham was, I am."

- That "I AM" statement? It's a direct reference to Exodus 3:14, where God tells Moses "I AM WHO I AM."

- The Jews listening didn't miss it. They picked up stones. Because to them, this was blasphemy—Jesus was claiming to be God Himself.

John 10:30 – "I and the Father Are One"

- Short. Direct. Impossible to misunderstand.

- "I and the Father are one."

- Not just in purpose or mission—Jesus was saying He was one in essence with God.

- And again, the people didn't react with a polite theological debate—they tried to kill Him (John 10:31-33).

Why? Because they got it. He was claiming divinity.

John 1:1, 14 – "The Word Was God... and Became Flesh"

- "In the beginning was the Word, and the Word was with God, and the Word was God. And the Word became flesh and dwelt among us."

- Translation? Jesus (the Word) existed *before everything*, was distinct from the Father (*with* God), and yet was fully divine (*was* God).

- And that "became flesh" part? That's the incarnation—God stepping into human form.

This is huge because it means Jesus isn't just *like* God. He *is* God.

Colossians 2:9 – "The whole Fullness of Deity in dwells Bodily"

- Paul doesn't mince words: "For in him the whole fullness of deity dwells bodily,"

- Not *part* of God. Not *a representation* of God.

- **All** of God's fullness was in Jesus. Period.

(That's a big deal.)

Titus 2:13 – "Our great God and Savior Jesus Christ,"

- This one's straightforward: Jesus isn't just our Savior. He's God.

- Paul ties them together with no hesitation—meaning the earliest Christians had no doubt about who Jesus was.

Hebrews 1:3 – "The exact imprint of God's nature"

- "He (Jesus) is the radiance of the glory of God and the exact imprint of his nature and he upholds the universe by the word of his power. ."

- If you want to know what God is like—look at Jesus.

- He isn't just *close* to God. He's the exact imprint of God's nature.

Philippians 2:6-7 – Jesus, God in Nature, Humbled for Us

- "Who, though he was in the form of God, did not count equality with God a thing to be grasped,"

- This early Christian hymn (yep, people were singing about Jesus being God *way* before church councils got involved) makes it clear:

  Jesus was God by nature.

  But instead of *clinging* to that status, He chose to step into our world as a human.

  His humility didn't make Him *less* God—it made His love for us even more incredible.

Revelation 1:8, 22:13 – Jesus Calls Himself "Alpha and Omega"

- In Isaiah 44:6, God declares, "I am the first and the last."

- In Revelation 1:8 and 22:13, Jesus says, "I am the Alpha and the Omega."

- Which means? Jesus and God are one.

Wait—Did Jesus Accept Worship?

Now, this is big. In Jewish belief, worship is for God alone—no exceptions.

- **Deuteronomy 6:13** "Fear(Worship) the Lord your God and serve Him only."

- **Exodus 34:14** "For you shall worship no other god."

So, if Jesus wasn't divine, He would've shut down any attempt to worship Him *immediately*. But He didn't. In fact, He accepted it.

Let's look at the receipts:

- **Matthew 14:33** – After Jesus walks on water (casual flex), the disciples worship Him:

    > And those in the boat worshiped him, saying, "Truly you are the Son of God.'"

- **John 20:28** – Doubting Thomas (or better needs more proof Thomas) gets his proof and calls Jesus *"My Lord and my God!"*

    Jesus doesn't correct him. Doesn't redirect him. He affirms it.

- **Matthew 28:9, 28:17** – After the resurrection, His disciples worship Him.

If Jesus wasn't God, accepting worship would've been *blasphemous*. But instead of stopping people, He welcomed it.

Why?

Because He is God. And He knew it.

It's Not Even Close—The Overwhelming Evidence for Jesus' Divinity

If you're still on the fence about Jesus being God, let's lay it all out. The New Testament isn't subtle about this. Jesus' divine **titles, actions, and the way He was treated** all point to the same conclusion—He was fully God in human form.

And remember—this wasn't happening in some loose, anything-goes religious landscape. **This was first-century Jewish culture. Monotheistic to the core. Worshiping anyone but God? That was a death sentence.**

And yet, **Jesus' followers worshiped Him anyway**. Because they knew.

But Wait, There's More!

Witnesses and Miracles (Or, "Things You Just Can't Fake")

Miracles weren't just nice party tricks. They weren't for show. Every miracle Jesus performed proved His divine authority—power over nature, sickness, demons, and even death. And these weren't private, backroom events. People saw them. Lots of people. Even His critics couldn't deny them.

Power Over Nature

- **Calming the storm** (Matthew 8:23-27, Mark 4:35-41, Luke 8:22-25)

  Jesus speaks. The wind and waves obey.

  His disciples? Shook. "What kind of man is this?" (Hint: Not just a man.)

  In Jewish thought, only God controls the sea (Psalm 89:8-9, Job 38:8-11).

- **Feeding the 5,000** (Matthew 14:13-21, John 6:1-14)

  Thousands of people. Five loaves. Two fish. Everyone eats.

  In John 6:35, Jesus calls Himself the "Bread of Life"—directly linking to God providing manna in the wilderness.

- **Walking on Water** (Matthew 14:22-33, Mark 6:45-52, John 6:16-21)

  Jesus defies the laws of physics.

  The disciples' response? They worship Him and say, "Truly you are the Son of God."

Power Over Disease

- **Healing the blind, the lame, and lepers** (John 9, Matthew 8:1-4, Mark 2:1-12)

  The prophets healed *through* God. Jesus heals by His own authority.

  He fulfills Isaiah 35:5-6—a prophecy that *only* the Messiah could fulfill.

- **Healing by His word alone** (John 4:46-54, Matthew 8:5-13)

  No touch, no ritual, no prep work—just a command.

Prophets prayed for healing. Jesus *commanded* it.

Authority Over Demons

- **Exorcisms** (Mark 1:21-28, Matthew 8:28-34, Luke 11:14-22)

    Jewish exorcists had to invoke God's name to cast out demons.

    Jesus? He just tells them to leave. And they do.

    Luke 11:20—"If it is by the finger of God that I cast out demons, then the kingdom of God has come upon you." (Translation: I *am* the kingdom of God.)

Power Over Death (And Not Just His Own)

- **Raising Jairus' Daughter** (Mark 5:35-43, Luke 8:40-56)

    A little girl is gone. Jesus tells her to wake up. She does.

- **Raising the Widow's Son** (Luke 7:11-17)

    The scene? A funeral procession. Jesus interrupts it with resurrection.

- **Raising Lazarus** (John 11:1-44)

    Four days dead. Jesus calls him out of the tomb.

    Then, He makes *this* claim: "I am the resurrection and the life."

    Not "I know about resurrection." I *am* resurrection.

Jesus didn't just do what the prophets did. He did what only God can do.

Forgiving Sins—Something Only God Can Do

Okay, this is a big one. In Jewish theology, only God can forgive sins. Period. And yet... Jesus keeps doing it.

- **Healing the paralytic** (Mark 2:1-12, Matthew 9:1-8, Luke 5:17-26)

    > Jesus tells the man, "Your sins are forgiven."

    > The Pharisees freak out: "Who can forgive sins but God alone?"

    > Jesus backs it up by healing the guy *on the spot*.

- **Forgiving the sinful woman** (Luke 7:36-50)

    > "Your sins are forgiven."

    > Again, people are shocked. "Who is this who even forgives sins?"

If Jesus were just a teacher, a prophet, or even an angel, He wouldn't have dared to say these things. But He does. And He backs it up.

The Resurrection—The Ultimate Proof

We'll cover this in full later (because it deserves its own deep dive), but let's be clear:

- If the resurrection didn't happen, Christianity falls apart.

- If it did happen, then Jesus isn't just important. He's God.

The resurrection proves:

- His claims were true.

- Sin and death were defeated.

- His divine identity is *undeniable*.

And this wasn't just some vague, spiritual "resurrection in our hearts." It was a real, bodily resurrection—witnessed by hundreds of people who were so convinced they were willing to die rather than deny it.

Final Thoughts

The debate about Jesus' divinity isn't just some intellectual exercise. It's everything.

- If Jesus isn't God, Christianity is just another religion.
- If He *is*, then His words, His power, and His promises are everything we've been searching for.

And based on the overwhelming evidence—His claims, His miracles, His authority, and His resurrection—the conclusion is pretty clear:

Jesus wasn't just a good man. He was God in human form.

And that changes *everything*.

Chapter 7
Jesus was Mentally Ill

L et's get straight to it—was Jesus crazy? Sounds like a bold (maybe even blasphemous) question, right? But stick with me. People have been wrestling with this for centuries. Some say Jesus displayed behaviors that could be labeled today as symptoms of mental illness—claims of divinity, radical teachings, emotional outbursts. Others argue the opposite—that he was the epitome of wisdom, moral clarity, and emotional stability. So, let's dig in. We're going to look at Jesus' words and actions through a psychological lens, compare them to modern mental health criteria, explore alternative explanations, and, yes, bring in a theological perspective (because, let's be real, you can't talk about Jesus without talking theology).

Psychological Analysis

If we're going to examine Jesus' mental state, we need to look at what he actually did and said. Jesus was compassionate—he healed the sick, stood up for the marginalized, and forgave even his enemies. That doesn't scream "mentally unstable." But hold on—some critics argue that his claims of being the Son of God and predicting his own death and resurrection could be signs of delusions of grandeur. That's a fair question.

Now, delusions of grandeur are often associated with schizophrenia or bipolar disorder with psychotic features. A person experiencing this might believe they have supernatural abilities or an exaggerated sense of their importance. Jesus, however, wasn't just claiming divinity—he backed it up with miracles (healing the blind, calming storms, raising the

53

dead—no small feats). And more than that, his teachings weren't erratic or nonsensical; they were deep, structured, and incredibly influential. If he were delusional, you'd expect inconsistencies or confusion. Instead, you get parables filled with wisdom and a moral philosophy that continues to shape the world today.

And what about stress? Jesus faced a lot of it—persecution, betrayal, execution. He got angry (flipping tables in the temple, anyone?), and he experienced deep sorrow (see: Gethsemane). But does this make him mentally unstable? Not at all. Strong emotions in response to extreme situations aren't symptoms of illness; they're signs of a fully functioning, deeply feeling human being. He also showed emotional intelligence—he connected with outcasts, taught patience, and even forgave the very people who nailed him to a cross. That's next-level emotional stability.

Diagnostic Criteria

Now, let's play the psychologist. Could Jesus be diagnosed with a mental disorder using today's standards? Schizophrenia is characterized by hallucinations, disorganized thinking, and delusions. Some say Jesus's hearing God's voice or claiming divine status fits the bill. But in his cultural and religious context, divine encounters were expected, not unusual. If he were hallucinating, we'd also expect incoherence or a break from reality—neither of which show up in the Gospels.

What about bipolar disorder? People with bipolar disorder experience extreme mood swings—periods of intense energy followed by deep depression. Jesus did have high-energy moments (his public ministry) and sorrowful ones (his agony in the garden), but these weren't erratic or self-destructive. Instead, his actions were consistent and purposeful. He didn't spiral into despair or recklessness—he followed a mission with clarity and focus.

So, does Jesus meet the criteria for mental illness? Nope. His words were coherent, his actions were measured, and his influence was profound. He wasn't showing signs of a mind in disorder—he was showing signs of a mind on a mission.

Alternative Explanations

If Jesus wasn't mentally ill, then what explains his behavior? First, let's talk cultural context. First-century Judea wasn't exactly a quiet suburb with a Starbucks on every corner. It was a hotbed of political and religious tension. Prophets spoke in bold, visionary language, often making claims that, in a modern clinical setting, might raise eyebrows. Jesus fit within this tradition—he spoke in parables, used metaphors, and claimed a divine calling. That wasn't bizarre for his time; it was expected.

Then there's the historical testimony. Jesus' followers—including highly educated folks like Paul—didn't see him as unhinged. Instead, they found his teachings to be transformative and life-changing. Even non-Christian historians like Josephus acknowledged his influence. And let's not forget: people who are truly mentally unstable don't typically start movements that last for over two thousand years.

Oh, and those miracles? Whether you believe them literally or not, they were reported by multiple sources. If he were simply delusional, you'd expect a lack of credibility. Instead, we see consistent testimonies of his supernatural actions and profound wisdom.

Theological Interpretation

Alright, let's bring in the big picture. Christianity teaches that Jesus was both fully God and fully man (John 1:14). If that's true, then his behavior wasn't just rational—it was divine. His miracles, his authoritative teachings, and his fulfillment of ancient prophecies point not to a mental disorder, but to a divine mission.

Take the Sermon on the Mount (Matthew 5-7). This wasn't the rambling of a confused or delusional man. It was a structured, ethical masterpiece that continues to challenge and inspire. His message of love, forgiveness, and justice wasn't erratic—it was revolutionary in the best possible way.

Then there's the big one—his resurrection. If Jesus was mentally ill and simply hallucinating about rising from the dead, how do we explain the empty tomb? How do we explain the sudden transformation of his followers from fearful deserters to bold preachers willing to die for their belief that he had conquered death? The theological perspective sees this not as delusion, but as divine truth breaking into history.

Long story short

So, did Jesus have a mental health disorder? Not even close. Let's get real—people who are mentally ill don't produce the most influential moral teachings in history. They don't inspire billions. They don't live with unwavering purpose, coherence, and emotional stability in the face of extreme suffering. And they certainly don't change the course of civilization.

Every angle—psychological, historical, and theological— points to the same conclusion: Jesus was not insane. He was intensely rational, emotionally intelligent, and spiritually profound. His moral and intellectual consistency, his ability to inspire across centuries, and his enduring impact all stand in direct contradiction to the idea of mental instability.

So, if he wasn't mentally ill, that leaves us with an even bigger question—what if he was exactly who he claimed to be? What if his words weren't the product of delusion, but of truth? If that's the case, then we're left with something far more challenging than a psychological diagnosis. We're left with a choice. And that choice—who we believe Jesus to be—is the one that really matters.

Jesus Finally Exposed

Chapter 8

Jesus Didn't Say What We Think He Said

N ot gonna lie—figuring out exactly what Jesus said is tricky business. The guy didn't walk around with a tape recorder, and the Gospels weren't written in real time like live-tweeting a sermon. Instead, his words were remembered, retold, translated, and written down decades later. Some folks say this means we can't trust the Gospels to give us his actual words. Others argue that, despite all the passing down and retelling, what we have is still an accurate reflection of what Jesus taught. So, what's the deal? This chapter dives into both sides, looking at how historians sort fact from fiction, how the Gospel writers did their thing, and whether we can trust what's been recorded. Buckle up—this is going to be a fascinating ride.

Criteria of Authenticity

Historians aren't guessing when they try to figure out what Jesus really said. They use some pretty solid methods, and three big ones stand out. Let's break them down.

> **Multiple Attestation**: Again, if a saying of Jesus pops up in multiple independent sources—like Mark, Q (that mysterious lost source that Matthew and Luke supposedly used), and even non-canonical texts like the Gospel of Thomas—it's got a higher chance of being the real deal. Why? Because different people in different places wouldn't have made up the same saying. Examples? Oh, there are plenty:

The Kingdom of God: If Jesus had a favorite topic, this was it. You see this theme in all three Synoptic Gospels (Matthew, Mark, and Luke) and even in the Gospel of Thomas. Safe to say, Jesus was all about preaching the Kingdom.

The Parable of the Mustard Seed: Found in Mark 4:30-32, Matthew 13:31-32, Luke 13:18-19, and—surprise, surprise—the Gospel of Thomas. It's all about something small growing into something huge. Sound familiar? Yep, Jesus probably said it.

The Beatitudes: These get top billing in Matthew 5:3-12 and Luke 6:20-23. They don't match word-for-word, but they both capture the same radical idea—blessed are the underdogs. Multiple attestation at its finest.

Dissimilarity: Now, this one's a little counterintuitive. If a saying of Jesus doesn't match what Jews were saying at the time and doesn't quite fit what the early Church wanted to preach, then it's probably authentic. Why? Because no one would have made it up.

Jesus Calling God "Abba": This is a big one. Jews had deep reverence for God's name, but Jesus went and called him "Abba," an intimate, fatherly term (Mark 14:36). It didn't match Jewish norms, and early Christians didn't really keep using it the same way. Sounds like something Jesus actually said.

No Divorce Policy: This is in Mark 10:2-12, and it's tough. Even Moses allowed divorce, and later Christians (like Paul) made some allowances. But Jesus? He was all-in on lifelong marriage. If early Christians were tweaking Jesus' words to fit their needs, they probably would have softened this one. But they didn't. That tells us it's probably original.

Love Your Enemies: Loving your neighbor? Sure, that was standard. Loving your enemy? That was Jesus-level radical. You see this teaching in Matthew 5:43-48 and Luke 6:27-36, and it doesn't line up with the Jewish or early Christian thinking of the time. No one had a reason to invent it, so odds are, Jesus really said it.

Embarrassment: As I said before, if a saying or event is awkward or problematic for the early Church, then it's probably true. Why? Because if they were making stuff up, they wouldn't include the embarrassing parts.

Jesus Gets Baptized by John: If Jesus was sinless, why would he need baptism? That's the kind of question that made early Christians squirm. But it's in Mark 1:9-11, Matthew 3:13-17, and Luke 3:21-22. The best explanation? It actually happened.

Jesus' Cry on the Cross: "My God, my God, why have you forsaken me?" (Mark 15:34). Not exactly the kind of thing you'd make up if you were trying to present a confident, victorious Messiah. The fact that it's included tells us it probably happened just like that.

Jesus Admits He Doesn't Know Everything: Mark 13:32 has Jesus saying even he doesn't know when the end of the world is coming. If you're trying to emphasize Jesus' divine knowledge, this is not the thing you'd put in his mouth. Which makes it all the more likely that he actually said it.

Comparative Analysis

If Jesus' words were passed down with care, why do the Gospels sometimes tell the same story differently? Good question. Here's why:

Oral Tradition Was Flexible: In an oral culture, stories were meant to be repeated, but word-for-word accuracy wasn't the goal. The meaning stayed the same, but the phrasing could shift. That's why Matthew and Luke's Beatitudes don't match exactly.

Each Gospel Had a Different Audience:

Matthew was writing to Jewish Christians, so he made sure to highlight Jesus as the fulfillment of Jewish prophecy.

Luke was all about inclusivity, focusing on Jesus' concern for the outcasts.

John—well, John went full-on theological, emphasizing Jesus as divine.

Different Sources, Different Details: The Synoptic Gospels share a lot because they used similar sources, but John? He wrote independently. That's why you see some totally unique sayings in John that don't show up elsewhere.

Translation Complications: Jesus spoke Aramaic. The Gospels were written in Greek. Some things might have been tweaked slightly in translation.

Editing for Effect: The Gospel writers weren't just reporters; they were storytellers with a message.

Sometimes they adjusted the wording to make that message clearer to their readers.

Hmmm?

So, did Jesus really say what we think he said? Short answer: Yes. Long answer: The Gospels weren't dictated transcripts, but the words we have were carefully preserved through oral tradition, written sources, and the testimony of those who followed him. The methods historians use—multiple attestation, dissimilarity, and embarrassment—consistently show that Jesus' sayings weren't fabricated or heavily altered. Sure, the Gospel writers had their own styles and audiences, but that doesn't mean they made stuff up. They were preserving what Jesus taught, making sure his words would last. And guess what? They did. We're still talking about them today. That alone tells you something.

Jesus Finally Exposed

Chapter 9

Jesus isn't Lord

Jesus isn't LORD? Let's not rush past this. It's easy to ask, *Can we prove Jesus isn't Lord?* But maybe that's the wrong door to knock on.

Think of it like this— Astronomers don't *see* black holes directly. You know why? Because black holes are, well... black. In space. Which is also black. It's like looking for a shadow in the dark. Oh, also, they are holes.

So instead of staring into the void, they look around it. For clues. They watch how stars behave nearby, how gravity bends light, how movement shifts. And when enough signs align, they say: *That's where the black hole is.*

Now, maybe that's how we approach this question too.

Let's not just try to *disprove* the Lordship of Jesus.
Let's look around.
Let's trace the evidence.
Let's ask: Is Jesus... LORD?

What Does "Lord" Even Mean?

It's an old word. Ancient, really.

Once, "lord" meant someone with power. A landowner. A noble. A ruler. Someone in charge. Someone others followed. Someone whose voice changed things.

Even now, when we say "lord," we're talking about influence. Authority. Weight.
A voice that moves something in us.

But when we talk about *Jesus* as Lord—we're not just giving Him a fancy title.

He's not Lord of the Manor. Not Lord Jesus Fauntleroy. This isn't English nobility.

This is something far greater.

Belief in Jesus as LORD

Messianic Expectations and Fulfillment of Prophecies

Many Jews of the time were expecting a Messiah, though interpretations varied. Jesus' followers claimed he fulfilled Old Testament prophecies about the suffering servant (Isaiah 53) and the Davidic King (2 Samuel 7:12-16).

Many Jews rejected Jesus because he did not fulfill expectations of a military or political Messiah who would overthrow Roman rule and restore Israel's sovereignty.

Testimony of Eyewitnesses and Early Christian Growth

The early Christian movement grew rapidly despite severe persecution. Key witnesses, such as the apostles, claimed to have seen the resurrected Jesus and were willing to suffer and die rather than renounce their testimony. This resilience and conviction convinced many that their belief in Jesus' LORDship was based on real experiences rather than fabrication.

One of the strongest arguments for Jesus' LORDship comes from the testimony of those who claimed to have encountered him, both before and after his death. The earliest followers of Jesus, including his disciples and other eyewitnesses, played a crucial role in the rapid expansion of Christianity, despite intense persecution and opposition. Their unwavering conviction and willingness to suffer and die for their belief suggest they genuinely believed Jesus was LORD.

And testimonies from people who initially doubted or opposed Jesus. For instance, Thomas (John 20:24-29) doubted the resurrection until he physically touched Jesus' wounds, and James, Jesus' own brother, was skeptical of Jesus' ministry before becoming a key leader in the early church (Mark 3:21; Acts 15:13-21). Paul, formerly Saul of Tarsus, was a zealous persecutor of Christians until he had a dramatic encounter with the risen Christ on the road to Damascus (Acts 9:1-19). The transformation of such individuals adds credibility to their testimony.

The Martyrdom of the Apostles

A striking feature of the early Christian movement is the willingness of its leaders to endure suffering, imprisonment, and even death rather than recant their belief in Jesus as LORD. According to historical traditions:

Peter was crucified upside-down in Rome.

James, the brother of John, was executed by King Herod Agrippa I (Acts 12:2).

Paul was beheaded in Rome.

Thomas was reportedly speared to death in India.

James, the brother of Jesus, was stoned to death by Jewish authorities.

Unlike many religions of the time, Christianity appealed to Jews and Gentiles, slaves and free, rich and poor, breaking social and ethnic barriers.

Across different regions and cultures, Christian teachings remained largely consistent, indicating that early believers did not invent or modify the core message of Jesus' LORDship.

Non-Christian sources confirm the rapid expansion of Christianity and the unyielding faith of its followers. Roman historian Tacitus (c. 116 AD) describes how Christians were persecuted under Emperor Nero and states that they worshipped Christ as a divine figure. Pliny the Younger (c.

112 AD) wrote to Emperor Trajan about Christians in Bithynia, describing how they gathered regularly, worshiped Jesus as a god, and refused to renounce their faith despite threats of execution.

Roman and Jewish Historical Sources

Both Jewish leaders and Roman officials rejected Jesus' claims of authority. Jewish leaders saw him as a blasphemer, and Romans saw him as a potential insurrectionist but not divine. This official opposition cast doubt on his identity as LORD in the eyes of many.

Non-Christian sources, such as Tacitus, Josephus, and Pliny the Younger, confirm Jesus' existence. They demonstrate that Jesus was a historical figure whose followers worshiped him as divine, lending credibility to the claim of his LORDship.

He Wasn't Just a Lord. He was The LORD.

When Jesus walked the earth, some expected a new king. A deliverer. A political leader who would set things right.

But Jesus didn't just bring a new agenda.
He brought a new revelation.

He used the name no one used lightly—*Yahweh.*
The highest name for God.
"I AM."   The One who is.   The One who brings everything into being
And that's the name the English Bible translates as LORD— in all caps.

So, when we say "Jesus is LORD," we're saying something massive.
We're saying He's not just from God.
He *is* God.   He's divine.   He's the LORD of lords.

Let's pull this together

The question of Jesus' LORDship remains central to Christian faith and one of the most debated theological issues in history. The historical evidence overwhelmingly supports the claim that Jesus existed, was crucified, and that he rose from the dead. His fulfillment of messianic prophecies, the testimony of his disciples, and the rapid and unwavering growth of early Christianity—despite severe persecution—point to his divine nature. Jesus himself made claims that only God could make, such as the authority to forgive sins (Mark 2:5-7) and his eternal existence (John 8:58), which led to charges of blasphemy from religious authorities.

And if it's really true—If Jesus really is LORD—Then everything changes.

If HE is LORD then He is The One who brings everything into being. He is your creator.

Other religions reject Jesus' LORDship because affirming it would undermine their core beliefs. Judaism denies Jesus as the Messiah because his mission did not align with traditional expectations of a political redeemer. However, the suffering servant prophecy in Isaiah 53 closely matches Jesus' life and mission, challenging the Jewish position. Islam's denial of Jesus' divinity is based on the belief in strict monotheism, yet Jesus' unique miracles, virgin birth, and sinless nature (acknowledged in the Quran) set him apart from all other prophets, aligning with the Christian claim of his divine nature and being LORD. Skeptics who see Jesus only as a moral teacher fail to account for his extraordinary claims of authority, resurrection, and the profound impact he had on history.

Not Just a Title—A Surrender

In Romans 10:9–10, Paul, a convert to Jesus, writes: "If you declare with your mouth, 'Jesus is LORD,' and believe in your heart that God raised him from the dead, you will be saved."

To declare "Jesus is LORD" was more than a spiritual phrase in the early church. It was a courageous act. In a world where Caesar was hailed as lord, choosing Jesus instead meant living under a different rule—even if it cost you.

Today, the pressures may look different, but the heart of that decision hasn't changed.

Calling Jesus LORD is not just giving Him a name—it's giving Him authority. It's choosing His way over our own. It's saying, Jesus, You lead—I'll follow.

LORD of My Everyday

So, what does it look like when Jesus is truly LORD?

He leads how I treat people. I don't get to cling to bitterness or hold onto pride. He says forgive, so I do.

He guides how I handle pressure. When I want to give in to fear or control, He reminds me to trust Him.

He defines my values. Even when culture pushes one way, I listen for His voice.

And that's the thing about LORDship—it's personal. It's not just about admiring Jesus from a distance. It's about handing Him the keys to every room of my life.

LORD of Lords

Scripture doesn't just call Him LORD. It calls Him LORD of Lords.

That means above every voice, every influence, every pull in your heart—Jesus stands higher.

And even when His way feels upside-down—when He says, "Lose your life to find it" or "The greatest is the servant"—if He really is LORD, then He's worth trusting. With everything.

The gavel comes down, and the verdict is...

Ultimately, the evidence points to Jesus as LORD—the crucified and risen Son of God. His LORDship is not merely a theological doctrine but a call to faith, redemption, and eternal life for those who accept him. The significance of this claim is why the debate endures, and why belief in Jesus as LORD remains the cornerstone of Christian faith.

Chapter 10

Jesus didn't rise from the dead

W hoa, let's get real for a second. The resurrection of Jesus—one of the biggest claims in human history. Either it happened, or it didn't. And depending on which side you're on, this changes everything.

Christians say Jesus was crucified, buried, and then—three days later—boom. Back to life. But skeptics? They're not so sure. Maybe it was a misunderstanding, a trick of the light, or something that just got exaggerated over time.

So, what do we do with this? Well, we dig in. We look at the evidence, the counterarguments, and—if we're feeling bold—our own assumptions. Let's walk through it together. (No pressure, but this might be the most important investigation you ever do.)

The Skeptics Speak: Is the Resurrection Just a Story?

Some folks argue that the resurrection is more of a legend than an actual event. And their reasons? Not totally unreasonable. Let's take a minute and step into their shoes. If you're someone who thinks the resurrection story is more fairy tale than fact, here's what you might be thinking:

- **The Gospels Were Written After the Fact.** The main sources we have—Matthew, Mark, Luke, and John—were written decades after Jesus lived. That's a decent amount of time for things to get... let's say, *embellished.* Ever played the telephone game? A simple phrase can turn into something completely different after it's passed through a few people. Now, imagine an event being passed down orally for 30, 40, even 50 years before being written down. Could details have changed? Could people have filled in gaps with what they *wanted* to believe? Skeptics say, absolutely.

- **Not a Lot of External Proof.** Sure, we've got the Bible. But what about other sources? The big historians of the time—Josephus, Tacitus—mention Jesus, sure, but they don't exactly confirm that He rose from the dead. And that's kind of a big deal. If this was *the* event of the century (or, you know, of *all time*), wouldn't we expect a few more neutral voices chiming in? Maybe a Roman guard's diary entry? A government record? Even a note from an annoyed local saying, "This Jesus guy won't stay in His grave!" But nope. Nothing quite like that. And for skeptics, that's a problem.

- **People Wanted It to Be True.** Have you ever wanted something so badly that your mind started to convince you it was real? (Think of that text you *thought* you saw from your crush… but nope, it was just an email from your dentist.) Some argue that the disciples were dealing with something similar. They had left *everything* to follow Jesus. Then, suddenly—He was gone. Dead. Executed like a criminal. That kind of heartbreak doesn't just go away. Maybe, just maybe, in their grief, their hope, and their desperation, they *believed* He was alive again—even if He wasn't.

Now, let's be honest—these are fair points. The lack of neutral, contemporary sources makes some people hesitant to take the resurrection at face value. And honestly? I get it. If someone told me a guy had come back from the dead last week, I'd need some serious proof. (Like, *show me the guy, let me shake his hand, and let's grab coffee together* level of proof.)

So, if you've ever wrestled with doubts about the resurrection—congrats! You're asking the same questions people have been asking for over 2,000 years. But hold on, because the other side of the argument? It's got some things to say, too...

But Then There's This...

Let's be honest—skeptics bring up some solid points. People don't just rise from the dead. Science, experience, common sense—none of it backs that up.

But those who believe in the resurrection? They've got some serious responses. And they don't just come from blind faith or wishful thinking. They come from history, from logic, from people who, quite frankly, had no reason to lie.

Let's walk through a few:

Early Testimonies

Before the Gospels were even written, Paul was talking about Jesus' resurrection in his letters. And Paul? Not exactly a guy looking to create a feel-good movement. He started as a hardline enemy of Christianity, a man who arrested and persecuted believers. Then, suddenly, he flipped—claiming he had an encounter with the risen Jesus. The turnaround was so extreme that even non-Christian historians take notice.

Take **Bart Ehrman**, a well-known agnostic scholar. He doesn't believe in the resurrection, but he does acknowledge that Paul truly believed Jesus appeared to him. That's significant. Why would Paul, a man so committed to wiping out Christianity, suddenly become one of its greatest advocates?

The 500 Witnesses

Paul also mentions in 1 Corinthians 15:6 that Jesus appeared to over 500 people at once. That's a big number. And the thing is, Paul wasn't writing this centuries later when memories got fuzzy. He was writing when people who were there could still say, "Wait a minute, I was in Jerusalem that day, and that didn't happen."

It wasn't just Christians talking about Jesus appearing after His death. The Roman historian Tacitus, who had no reason to push a pro-Christian agenda, acknowledged that Jesus was executed under Pontius Pilate—and that His followers believed He had risen. Josephus, a Jewish historian, also recorded that early Christians claimed Jesus appeared to them after His death.

Even if you don't believe in the resurrection, you have to wrestle with the fact that people at that time believed it happened—so much so that they were willing to die for it.

The Women at the Tomb

Now, if you were making up a story in ancient Jewish culture, you wouldn't choose women as your key witnesses. Back then, women's testimony wasn't considered strong in court. Yet, the Gospels say women were the first to find the empty tomb and see Jesus alive.

If early Christians were inventing this story, why would they include a detail that actually hurts their credibility in that culture? The only reason you'd keep that in? Because it actually happened.

And Then There's the Unlikely Converts

The resurrection didn't just convince Jesus' followers—it turned skeptics into believers.

- **James, Jesus' brother**, wasn't a follower of Jesus during His ministry. But after the resurrection? He became a leader in the early church and was eventually martyred. What changed?

- **The historian Pinchas Lapide, a Jewish scholar**, doesn't believe in Christianity but has written that the resurrection of Jesus is the best explanation for how the Christian movement exploded overnight. A non-Christian saying the resurrection is historical? That's not nothing.

The Empty Tomb: Coincidence, Conspiracy, or...?

Okay, let's talk about this. It's one of the biggest questions in history. The tomb was empty. That much is generally accepted. What's debated is why.

Skeptics have thrown out several explanations, trying to make sense of it without jumping to the whole "resurrection" thing. Let's break down the main ones:

Jesus Didn't Actually Die.

This is known as the "swoon theory." The idea? Jesus didn't really die on the cross—He just *passed out* from exhaustion and blood loss. Then, a few days later, He woke up, somehow rolled away a massive stone, snuck past the Roman guards, and strolled out like nothing happened.

Problem? The Romans were *experts* at killing people. Crucifixion wasn't just a slap on the wrist—it was a brutal, public execution. These guys weren't rookies. They even stabbed Him in the side with a spear to *double-check* that He was dead. And let's be real: if He *had* somehow survived, He wouldn't have walked out looking like a victorious, risen Savior—He would have looked like He needed an ambulance.

The Wrong Tomb Theory: A Swing and a Miss

Alright, so let's talk about one of the more *creative* explanations for the resurrection: the Wrong Tomb Theory. The idea? That early Sunday morning, the women who went to Jesus' tomb—grieving, exhausted, maybe still wiping tears away—accidentally went to the wrong one.

And, surprise! It was empty.

They freaked out. Ran back. Told the disciples. And boom— Christianity was born… all because of a *GPS fail*.

Now, let's just sit with that for a second. Because on the surface, it sounds almost *plausible*, right? I mean, I've lost my car in a parking lot before. Maybe they just got mixed up.

But here's the problem—actually, several problems.

First, Joseph of Arimathea's tomb wasn't just some random cave in a field. This was a well-known, easily identifiable burial site owned by a prominent figure. If you're going to bury the most controversial, talked-about person in town, you don't just casually toss him in an unmarked grave and hope for the best.

Second, the Jewish and Roman authorities had one job: keep Jesus *in* the tomb. This wasn't some forgotten grave—this was under watch. And let's be honest, if the disciples started preaching that Jesus had risen from the dead, don't you think the authorities would've marched down to the correct tomb, pointed at the body, and shut the whole thing down?

And finally—the disciples weren't just banking their lives on an *oops* moment. These guys didn't just believe the tomb was empty. They believed they *saw* Jesus—talked to Him, touched Him, ate with Him. And they didn't spread that message with "Well, we're *pretty sure* we had the right tomb, but hey, who knows?" energy. They *died* for this.

So yeah, the *Wrong Tomb Theory*? It's a stretch. A pretty desperate one, at that. Because at the end of the day, the real issue isn't whether the women got lost.

It's whether Jesus actually walked out of that tomb alive.

And if He did—well, that changes everything.

The Body Was Stolen.

Another popular theory is that someone took the body. Maybe the disciples snuck in and grabbed it. Sounds simple, right? Until you consider a few details:

- There were Roman guards posted at the tomb. And these weren't rent-a-cops. These were well-trained soldiers whose job was to keep things secure.

- The tomb was sealed with a massive stone. We're talking about something *several men* would have struggled to move.

- The disciples—who, by the way, were terrified and hiding after Jesus' death—suddenly turned into fearless preachers, willing to be imprisoned, tortured, and executed for proclaiming He had risen. Now ask yourself: would *you* be willing to die for something you *knew* was a hoax?

Even skeptics struggle with this. Take Chuck Colson, the former White House lawyer involved in the Watergate scandal. He wasn't a religious guy. But after his own experience with a cover-up, he said:

"I know the resurrection is a fact, and Watergate proved it to me. How? Because 12 men testified they had seen Jesus raised from the dead, then they proclaimed that truth for 40 years, never once denying it. Every one of them was beaten, tortured, stoned, and put in prison. They would not have endured that if it weren't true. Watergate embroiled 12 of the most powerful men in the world—and *we* couldn't keep a lie for three weeks."

If a group of powerful politicians *couldn't* keep a lie together, how could a bunch of fishermen and tax collectors?

Mass Hallucination?

Okay, maybe Jesus' followers just *thought* they saw Him. Maybe they were grieving so hard that their minds played tricks on them.

Here's the issue: hallucinations are personal experiences. They don't happen in big, shared groups. Yet, the Bible records that over 500 people saw Jesus alive after His death. And not just His followers—skeptics, too.

One of the most famous was Paul (formerly Saul), a guy who *hated* Christians. He made a career out of arresting them. But then? He had an encounter with the risen Jesus—and it changed everything. He went from *persecutor* to *preacher* and ended up dying for the very message he once tried to destroy.

The Transformation of the Disciples: What Changed?

Let's talk about the disciples for a second. I've said some of this before but I need to recap. These guys weren't superheroes. They weren't fearless warriors or polished public speakers. They were fishermen, tax collectors, and everyday people. And when Jesus was arrested? They scattered.

Peter—the guy who swore up and down he'd never leave Jesus—denied knowing Him three times. Three. And the others? Hiding. Doors locked. Absolutely terrified.

But then something happened.

Because after the resurrection, these same men burst onto the scene like a completely different crew. Suddenly, they were in the streets, publicly preaching about Jesus—right in front of the very people who had Him crucified. And when the authorities arrested them, threatened them, beat them, even killed them—they still wouldn't stop.

So, here's the question: *What changed?*

Remember these guys from Chapter 9. Let's look at the cost of their message:

- **Peter?** Crucified upside down.

- **James?** Stoned to death.

- **Paul?** Beheaded in Rome.

- **Thomas?** Speared to death in India.

- **Andrew?** Crucified.

- **John?** Exiled after being boiled alive (the only one who died of old age).

Now, sure—people die for lies all the time. That's not new. But here's the key difference:

People don't willingly die for something they *know* is a lie.

If the disciples had *stolen* the body, or *made up* the resurrection, at least *one* of them would have cracked under pressure. Someone would've confessed. Someone would've slipped up. Yet every single one of them held firm—even when it cost them everything.

That's not normal. That's not psychology. That's conviction.

Even skeptics recognize this is a problem. Blaise Pascal, the 17th-century mathematician and philosopher, put it this way:

"I believe those witnesses that get their throats cut."

They weren't dying for an idea. They weren't dying for a philosophy. They were dying for something they claimed to have seen with their own eyes.

What Do Non-Christians Say?

Okay, maybe you're thinking, *"Well, that's what Christians believe. But what do outside sources say?"*

Good question. Let's check the history books.

Gerd Lüdemann, a well-known atheist New Testament scholar, says this about the disciples:

"It may be taken as historically certain that Peter and the disciples had experiences after Jesus' death in which Jesus appeared to them as the risen Christ."

Did you catch that? A non-Christian historian saying it's *historically certain* that the disciples believed they saw Jesus alive again.

Even **Bart Ehrman**, one of the world's leading agnostic Bible scholars, admits:

"We can say with some confidence that some of Jesus' disciples claimed to have seen Him alive after His crucifixion... This is, in my mind, a historical fact."

That doesn't mean they believe the resurrection actually happened. But it does mean the disciples weren't making it up. They *believed* they saw Jesus. And that belief turned a group of fearful deserters into fearless preachers willing to die for their message.

So, we're left with the big question: *What did they see?*

If it was a hoax, it would have unraveled. If it was a lie, someone would have cracked. If it was a hallucination, it wouldn't have been a shared experience.

But if it was real?

Well... then maybe it explains why these men changed the world.

The Resurrection – Hoax or History?

So, here we are. After all the theories, the counterarguments, and the deep dives into history, we're left with the same question people have been asking for 2,000 years: *Did Jesus really rise from the dead?*

Here's what we know. The resurrection of Jesus isn't just some vague religious claim—it's a historical event that reshaped the world. And when you examine the evidence—the eyewitness accounts, the radical transformation of Jesus' followers, the explosive growth of Christianity despite persecution—it becomes clear: something happened. And not just *something—something big.*

The Eyewitness Accounts: Too Many, Too Soon, Too Specific

One of the strongest pieces of evidence? The sheer number of people who claimed to have seen Jesus alive after His death. And not just one or two. We're talking multiple individuals, small groups, and even a crowd of 500 people at once (1 Corinthians 15:3-8). That's not the kind of detail you throw in if you're making up a story—especially when those people were still alive and could confirm or deny it.

And these weren't just brief, shadowy glimpses. The Gospel accounts describe Jesus talking, eating, and letting people touch Him. Thomas, the doubter, didn't just see Him—he put his hands in Jesus' wounds (John 20:27). That's not the kind of experience you confuse with a hallucination.

Even non-Christian sources take these claims seriously. **Gerd Lüdemann,** an atheist historian, states:

"It may be taken as historically certain that Peter and the disciples had experiences after Jesus' death in which Jesus appeared to them as the risen Christ."

Notice what he's saying. He doesn't *believe* in the resurrection, but he acknowledges that the disciples *believed* it—and they believed it so deeply that they went to their deaths rather than deny it.

The Disciples' Transformation: Fearful to Fearless

Before the crucifixion, the disciples were a mess. Peter denied Jesus three times. The others ran and hid. They weren't exactly brave.

But after the resurrection? Everything changed. Suddenly, these same men were boldly preaching in the streets, facing prison, torture, and execution—without backing down.

People don't willingly die for something they *know* is a lie. But the disciples were ready to do that and did. Something turned these men from cowards into warriors.

Even skeptics recognize this. Historian E.P. Sanders, who isn't a Christian, admits:

"That Jesus' followers (and later Paul) had resurrection experiences is, in my judgment, a fact."

Again—he's not saying he believes Jesus rose from the dead, but he acknowledges that the disciples *were convinced* they saw Him alive. And if they were lying? Well, hoaxes don't usually inspire people to endure beatings, stonings, and executions without *one* of them breaking.

The Empty Tomb: An Inconvenient Problem for Skeptics

Then there's the tomb.

Jesus was buried in a known, accessible location—a tomb owned by Joseph of Arimathea, a respected Jewish leader

(Matthew 27:57-60). That means anyone—Roman officials, Jewish leaders, skeptical onlookers—could go check it out.

And yet… no body was ever produced.

If the disciples had stolen it, the authorities could have exposed the fraud in minutes. If the Romans or Jewish leaders had taken it, they would have gladly rolled out the corpse to shut down the growing Christian movement. Instead? The best explanation they could come up with was a cover story—that the disciples stole the body (Matthew 28:11-15). But let's be honest—does it make sense that a group of terrified fishermen suddenly overpowered trained Roman guards and pulled off the greatest hoax in history?

And let's not forget: the first witnesses to the empty tomb were women (Matthew 28:1-10, Luke 24:1-12). That might not seem like a big deal now, but in the first century, a woman's testimony wasn't even admissible in court. If you were making up a story, you wouldn't choose women as your key witnesses. The fact that the Gospels include this detail actually strengthens their credibility—because no one would invent a story that made it harder to believe.

The Growth of Christianity: If It Was a Lie, Why Did It Spread?

Then there's this: Christianity didn't just survive—it exploded. And it did so under heavy persecution from both Jewish and Roman authorities.

If the resurrection was a myth, Christianity would have died out *quickly* under scrutiny. Instead, within a few decades, it spread across the Roman Empire. By the second century, Roman officials like Tacitus (Annals 15.44) and Pliny the Younger (Letters 10.96) were documenting the movement, baffled by its rapid growth.

People don't willingly suffer and die for something they *know* is a hoax. And yet, Christianity grew—not through military conquest, not through political power, but through the unshakable conviction of those who believed they had seen Jesus alive.

The Most Straightforward Explanation

According to Sir Arthur Conan Doyle's Sherlock Holmes, "When you have eliminated all which is impossible, then whatever remains, however improbable, must be the truth."

Here's where we land. The simplest explanation is the one Christians have believed for 2,000 years: **Jesus actually rose from the dead.**

Sure, it sounds wild. But let's be honest—history is full of things that sounded crazy at first. The earth revolving around the sun? People laughed at that. The idea that invisible germs cause disease? Ridiculous—until it wasn't.

So, is it so far-fetched to think that *if* there's a God who created life in the first place, He could bring someone back from the dead?

The empty tomb is a fact.

So, What Do You Do With That?

So, here we are. After all the debates, the theories, the counter-theories (and, let's be honest, the mental gymnastics some folks go through to explain it all away), we're left with one simple but massive question:

Did Jesus really rise from the dead?

And here's the thing—that's not just some feel-good, Sunday-morning, warm-and-fuzzy question. It's history-shaping. Life-altering. Reality-shifting.

Because the resurrection isn't just a story. It's not some myth cooked up by a few desperate followers trying to keep the movement alive. It's a claim backed by early sources, by people who saw something they couldn't unsee, by skeptics who became martyrs.

And let's be real—people don't die for something they know is a lie. They don't trade comfort for persecution, security for suffering, unless something shook them to their core.

Now, let's hit pause and be fair. Because some folks have tried to explain this away. Here's a quick recap of the alternative theories:

- **The disciples stole the body?** Then why die for a lie? That's not just a bad plan—it's an illogical one.

- **They hallucinated?** Group hallucinations aren't a thing. Science says so.

- **It was a myth that developed over time?** The resurrection was preached immediately, not centuries later. No time for legend to creep in.

- **Wrong tomb?** The authorities would've just pointed to the right one. Problem solved.

At some point, you have to ask: What's the simplest, most straightforward explanation?

For historian N.T. Wright, the answer is clear:

"As a historian, I cannot explain the rise of early Christianity unless Jesus rose again, leaving an empty tomb behind him."

This isn't just about faith—it's about evidence. And while belief in the resurrection is ultimately a personal decision, history presents a compelling case that something world-changing happened that first Easter morning.

The tomb was empty. The disciples were transformed. The movement spread.

So yeah, you could dismiss it. Call it a legend, wishful thinking, a theological fairy tale.

But if you do, you've got to answer this:

Why did people who had everything to lose stake their lives on it? And why, 2,000 years later, does an empty tomb still demand an answer?

And if the resurrection *is* true?

Well. That changes everything.

So, the real question isn't *did* it happen.

It's—*what are you going to do with it?*

Chapter 11

Now You Have to Decide

Okay. So now what?
Before this book, maybe you could say, *"I didn't know."* But now you do. Now you've got a decision to make. Where does Jesus stand in your life?

And really, it all boils down to two options:

- Jesus was exactly who He said He was—Son of God, Savior, Lord of Lords.

- Or... He wasn't.

That's it. No middle ground. No *"Well, He was just a good teacher."* He didn't leave that option on the table.

And honestly? No other leader in history has been as respected *and* made those kinds of claims. That's something.

Let's Recap:

- **The Bible?** Historically reliable. It holds up better than any other ancient text. Archaeology backs it. The manuscript evidence is overwhelming.

- **Jesus?** Not a liar. A liar wouldn't die for something they knew was false. And they certainly wouldn't inspire billions to follow.

- **Not a legend either.** His life and death are recorded by sources outside the Bible. No mythology here— just real history.

- **Not just a moral teacher**—Jesus spoke with authority, forgave sins, claimed divinity, and inspired followers who gave their lives believing he was the Son of God.

- **Was He mistaken about His identity?** Nope. His teachings were consistent. His prophecies came true. He fit the Old Testament predictions. He knew exactly who He was.

- **Did He claim to be divine?** Sure did. He forgave sins. He accepted worship. And His followers believed it so much they died rather than deny it.

- **Mentally ill?** No way. His teachings were too sharp, too deep. His impact too great. A delusional man doesn't launch a movement that lasts for thousands of years.

- **Jesus' Words Are Historically Reliable** – Multiple independent sources confirm His teachings, and historians recognize their authenticity.

- **Jesus Was Recognized as Lord by His Followers** – His disciples worshiped Him and refused to deny Him, even when faced with death.

- **The resurrection?** Not a hoax. Too many witnesses. Too many changed lives. Too much historical evidence.

So.

If Jesus *wasn't* who He claimed to be, well… that's that. You can close this book and move on. But if He *was*—if He really is the Son of God—shouldn't that mean something?

Think about it.

Does refusing to believe in freight trains stop one from hitting you if you're standing on the tracks? Belief doesn't change truth. Truth is truth.

And Jesus said:(John 14:6)

"I am the way, and the truth, and the life. No one comes to the Father except through me."

I read on a pen one time "If this simple pen was created, then maybe you have a Creator." Interesting thought, since we are so much more complex than a pen.

Three Questions for You:

1. If you were to die today, do you know—without a doubt—that you'd be with God in Heaven?

2. If you *did* die today, and God asked, *"Why should I let you into My Heaven?"* what would your answer be?

3. What are you doing with Jesus?

Number 1 is a big question.

Look, if you're thinking, *"I'm not sure,"* wouldn't you like to be?

If your answer to 2 was something like, *"I try to be a good person. I go to church. My mom's a Christian,"* I need to tell you—none of that saves you. Those are good things, but they're not *the thing*. No matter who you know, You have to decide for Yourself.

No amount of good works can earn it. Jesus already paid for it.

Jesus asked this: (Matthew 16:26)

"What will it profit a man if he gains the whole world and forfeits his soul?"

Many feel they will be a good person and become successful in this world and then decide when they are old, but do you really know when you will die and *forfeit your own soul*?

Jesus said, (Luke 19:10)

"For the Son of Man came to seek and to save the lost."

And:(John 5:24)

"Whoever hears my word and believes Him who sent me has eternal life. He does not come into judgment, but has passed from death to life."

Did you catch that? Jesus didn't say, *"Work harder. Get your act together. Follow all the rules."* Nope. He just said—*believe.* Salvation? It's a **free gift.** That's not how we usually think, but that's Jesus decided. The only thing you "do" is believe and accept the free gift.

1. **If Your Answer to the Third Question** (What are you doing with Jesus?) **Was** *"Nothing"*...

Then maybe it's time to do *something.*

Time for a reality check. The Bible says: (Romans 3:23)

"For all have sinned and fall short of the glory of God."

That's everyone. Me. You. All of us. Sin isn't just the big stuff—murder, theft, etc. It's the little stuff, too. Ever been impatient? Lied? Grumbled under your breath? Yep. Sin.

Do you realize you only need to steal once to be a thief, or murder once to be a murderer. So, if you've sinned only once (or a million times) you're a sinner.

And the glory of God - that's living forever in heaven.

Sin has a cost:(Romans 6:23)

"For the wages of sin is death."

Either you pay for it. Or you let Jesus pay it for you.

The good news? (Romans 5:8)

"God shows His love for us in that while we were still sinners, Christ died for us."

That's grace. That's love. That's Jesus.

And it's personal.

Jesus said (John 3:16) *"For God so loved the world..."*

Go ahead. Put your name in there. Let's modify it for just you.

For God so loved (your name) that He gave His one and only Son, that if I believe in Him I shall not perish but I will have eternal life.

Jesus called it a gate or door (Matthew 7:13-14, John 10:9)

"For the gate is wide and the way is easy that leads to destruction, and those who enter by it are many. For the gate is narrow and the way is hard that leads to life, and those who find it are few. I am the door. If anyone enters by me, he will be saved and will go in and out and find pasture (eternal life in heaven with Him)." There is only one way, not many ways, to get to Jesus.

Romans tells us: (Romans 10:9)

"If you confess with your mouth, 'Jesus is Lord,' and believe in your heart that God raised Him from the dead, you will be saved."

That's it. No hoops. No hurdles. Just faith.

A Simple Prayer

If you know you need Jesus, and you're ready to make that decision, you can pray this. Right now.

Lord Jesus, I know that I am a sinner. I believe You died for my sins and rose from the dead. I turn from my sins and ask for Your forgiveness. I invite You into my heart and life. Be my Savior and Lord. Teach me to follow You and learn how to make you my Lord. Amen

That's it..

Not complicated. Not a magic formula. Just an honest prayer from your heart.

And if you prayed that, and really meant it, You are saved.

Jesus said "There is joy before the angels of God over one sinner who repents."

Translation: Heaven just threw a party. For you.

What's Next?

- **If you know another Christian, tell them what you did!**

- Get a Bible. (The YouVersion app is a great place to start. Try the ESV—it's pretty easy to understand.)

- Be aware that Satan is not pleased, so expect him to throw some doubt your way.

- That's why it important to find a solid church. (Walk in. If you feel at home, you've probably found your place.)

- If you need help or have questions, call 1-888-537-8720. They can get you started.

Welcome To The Family!!!

Appendix
Is Heaven Real?

A Personal Reflection

Before we begin, let me offer a little nudge. If you've landed here early—maybe out of curiosity or eagerness—I encourage you to backtrack and begin from the start of this journey. Some things are best seen in order, like watching a sunrise slowly instead of flipping on the lights.

This chapter wasn't originally in the plan. But as I walked through everything we've talked about, it became clear: we needed to pause and talk about where all of this leads—what happens after we draw our last breath. And most importantly, *is Heaven real?*

Because if Jesus came to restore what was broken between us and God, then surely it's not just for this life alone. Surely, it's for what's beyond it.

What Happens After This Life?

People have always wondered what comes after death. And naturally, the world has come up with many answers—some thoughtful, others imaginative. Here are a few:

- **Cessation of Consciousness** – We stop existing. No more thoughts. No more "us."

- **Reincarnation** – We return to live again, as someone or something else.

- **Spirit World** – We enter a realm of spirits.

- **Unconscious Waiting** – We sleep until something else happens.

- **Simulation Theory** – This whole life is a digital illusion.

- **Energy Transformation** – Our soul becomes part of the universe's energy.

- **The Egg Theory** – We're all one soul, living out every life that's ever been lived.

They're interesting. Creative. Even comforting, to some degree. But here's the question—what are they based on? Because, honestly, none of those theories come with a passport stamp from someone who's actually been there.

All of them require belief. Faith.

Belief by Itself Isn't Enough

And that's the thing about belief—it's powerful, but it doesn't make something *true*. You can believe your house isn't on fire… but if the roof is collapsing, your belief won't change the reality. What's real is what's real, no matter how we feel about it.

So, if we're going to stake our lives—or our eternity—on something, wouldn't it be wise to make sure that belief has a solid foundation?

That's why the Bible matters. Not as a fallback or tradition, but as a reliable witness. Throughout this book, we've looked at the credibility of Scripture, the person of Jesus, and the remarkable consistency of His message. He wasn't just another teacher. He taught *as one with authority*. Not just quoting truth—*He was truth.*

Jesus Didn't Guess About Heaven—He Knew

If we believe Jesus is who He said He is—divine, risen, Lord of all—then we have to take seriously what He said about what's next. And Jesus talked a lot about Heaven. Roughly 15% of His recorded words pointed toward it.

Here are just a few of His promises:

> "Blessed are the poor in spirit, for theirs is the **kingdom of heaven**."
> – Matthew 5:3

> "Rejoice and be glad, for your reward is great **in heaven**…"
> – Matthew 5:12

> "In **my Father's house** are many rooms… I go to prepare a place for you." – John 14:2-3

> "I will give you the keys of the **kingdom of heaven**…"
> – Matthew 16:19

> "Unless you turn and become like children, you will never enter the **kingdom of heaven**."
> – Matthew 18:3

Jesus wasn't speculating. He was speaking as someone who had come *from* Heaven… and would return there.

> "No one has **ascended into heaven** except he who **descended from heaven**, the Son of Man."
> – John 3:13

> "I am **ascending to my Father** and your Father, to my God and your God."
> – John 20:17

The Bible tells us that He did ascend—eyewitnesses watched Him go.

> "He was lifted up, and a cloud took Him
> **out of their sight**."
> − Acts 1:9

And now?

> "He sat down at **the right hand of God**."
> − Mark 16:19

The writer of Hebrews says "Christ has entered... into heaven itself, now to appear in the presence of God on our behalf." − Hebrews 9:24

But What About Hell?

Jesus didn't shy away from hard truths. Just as Heaven is real, He warned us that Hell is too. It's not comfortable to talk about, but He did—often. Why? Because love tells the truth, even when it's hard to hear.

> "Whoever says, 'You fool!' will be liable
> **to the hell of fire**."
> − Matthew 5:22

> "Better to lose one part of your body than
> for your whole body to go **into hell**."
> − Matthew 5:29-30

> "They will throw them into the **fiery
> furnace... weeping and gnashing of
> teeth**."
> − Matthew 13:42

> "You serpents... how are you to escape
> being **sentenced to hell**?"
> − Matthew 23:33

> **"In Hades(Hell)...** the rich man cried out,
> 'I am in anguish in this flame.'"
> — Luke 16:23-24

If eternal joy exists, then the absence of that joy—eternal separation—must also exist. Jesus wasn't trying to scare us. He was showing us what's at stake. Because Heaven is a place prepared *for us*. And Hell is a place we were never meant to go.

So... What Do You Believe?

You and I both know—we can't prove Heaven or Hell in a lab. But we *can* listen to the One who's been there.

Jesus didn't just talk about life. He laid down His own so we could *have* life, forever. And now, He invites us to believe not just in a comforting theory, but in *Him*—the risen Savior, the way to the Father, the open door to eternity.

The decision is yours.

But make it with your eyes open. Make it with your heart honest. Make it knowing there is One who loves you enough to tell you the truth... and then give everything so you could walk with Him, forever.

Full Scripture Passages for the book, plus a few more

Chapter 1
The Bible Is Not Accurate

Walls of Jericho: Joshua 6:20 So the people shouted, and the trumpets were blown. As soon as the people heard the sound of the trumpet, the people shouted a great shout, and the wall fell down flat, so that the people went up into the city, every man straight before him, and they captured the city.

High Priest Calaphas: John 11:49-52 But one of them, Caiaphas, who was high priest that year, said to them, "You know nothing at all….

Pontius Pilate: Matthew 27:2 And they bound him and led him away and delivered him over to Pilate the governor.

House of David/King David: 1 Samuel 16:13, 2 Samuel 7:16 Then Samuel took the horn of oil and anointed him in the midst of his brothers. And the Spirit of the Lord rushed upon David from that day forward. And Samuel rose up and went to Ramah. — And your house and your kingdom shall be made sure forever before me. Your throne shall be established forever.

Pool of Siloam: John 9:7 and said to him, "Go, wash in the pool of Siloam" (which means Sent). So he went and washed and came back seeing.

Hezekiah Tunnel: 2 Chronicles 32:30 Hezekiah blocked the upper outlet of the spring of Gihon and channeled the water down to the west side of the City of David

Babylonian Siege of Jerusalem: Jeremiah 25:1-13 .... Therefore thus says the Lord of hosts: Because you have not obeyed my words, [9]behold, I will send for all the tribes of the north, declares the Lord, and for Nebuchadnezzar the king of Babylon, my servant, and I will bring them against this land and its inhabitants, and against all these surrounding nations. I will devote them to destruction, ...

Chapter 2
Jesus Was a Liar

John 7:8–10: "You go up to the feast. I am not going up to this feast, for my time has not yet fully come." After saying this, he remained in Galilee. But after his brothers had gone up to the feast, then he also went up, not publicly but in private."

Mark 4:10–12: "And when he was alone, those around him with the twelve asked him about the parables. And he said to them, 'To you has been given the secret of the kingdom of God, but for those outside everything is in parables, so that "they may indeed see but not perceive, and may indeed hear but not understand, lest they should turn and be forgiven.""

Matthew 24:34: "Truly, I say to you, this generation will not pass away until all these things take place."

John 14:6: "Jesus said to him, 'I am the way, and the truth, and the life. No one comes to the Father except through me.'"

John 18:37: "Then Pilate said to him, 'So you are a king?' Jesus answered, 'You say that I am a king. For this purpose I was born and for this purpose I have come into the world—to bear witness to the truth. Everyone who is of the truth listens to my voice.'"

Matthew 5:37: "Let what you say be simply 'Yes' or 'No'; anything more than this comes from evil."

Chapter 3
Jesus Was Just a Moral Leader

Matthew 5–7: The Sermon on the Mount:

Seeing the crowds, he went up on the mountain, and when he sat down, his disciples came to him. And he opened his mouth and taught them, saying:

Blessed are the poor in spirit, for theirs is the kingdom of heaven.

Blessed are those who mourn, for they shall be comforted.

Blessed are the meek, for they shall inherit the earth.

Blessed are those who hunger and thirst for righteousness, for they shall be satisfied.

Blessed are the merciful, for they shall receive mercy.

Blessed are the pure in heart, for they shall see God.

Blessed are the peacemakers, for they shall be called sons of God.

Blessed are those who are persecuted for righteousness' sake, for theirs is the kingdom of heaven.

Blessed are you when others revile you and persecute you and utter all kinds of evil against you falsely on my account. Rejoice and be glad, for your reward is great in heaven, for so they persecuted the prophets who were before you.

You are the salt of the earth, but if salt has lost its taste, how shall its saltiness be restored? It is no longer good for anything except to be thrown out and trampled under people's feet.

You are the light of the world. A city set on a hill cannot be hidden. Nor do people light a lamp and put it under a basket, but on a stand, and it gives light to all in the house. In the same way, let your light shine before others, so that they may see your good works and give glory to your Father who is in heaven.

Do not think that I have come to abolish the Law or the Prophets; I have not come to abolish them but to fulfill them. For truly, I say to you, until heaven and earth pass away, not an iota, not a dot, will pass from the Law until all is accomplished. Therefore whoever relaxes one of the least of these commandments and teaches others to do the same will be called least in the kingdom of heaven, but whoever does them and teaches them will be called great in the kingdom of heaven. For I tell you, unless your righteousness exceeds that of the scribes and Pharisees, you will never enter the kingdom of heaven.

You have heard that it was said to those of old, "You shall not murder; and whoever murders will be liable to judgment." But I say to you that everyone who is angry with his brother will be liable to judgment; whoever insults his brother will be liable to the council; and whoever says, "You fool!" will be liable to hell of fire. So if you are offering your gift at the altar and there remember that your brother has something against you, leave your gift there before the altar and go. First be reconciled to your brother, and then come and offer your gift.

Come to terms quickly with your accuser while you are going with him to court, lest your accuser hand you over to the judge, and the judge to the guard, and you be put in prison. Truly, I say to you, you will never get out until you have paid the last penny.

You have heard that it was said, "You shall not commit adultery." But I say to you that everyone who looks at a woman with lustful intent has already committed adultery with her in his heart. If your right eye causes you to sin, tear it out and throw it away. For it is better that you lose one of your members than that your whole body be thrown into hell. And if your right hand causes you to sin, cut it off and throw it away. For it is better that you lose one of your members than that your whole body go into hell.

It was also said, "Whoever divorces his wife, let him give her a certificate of divorce." But I say to you that everyone who divorces his wife, except on the ground of sexual immorality, makes her commit adultery. And whoever marries a divorced woman commits adultery.

Again you have heard that it was said to those of old, "You shall not swear falsely, but shall perform to the Lord what you have sworn." But I say to you, Do not take an oath at all, either by heaven, for it is the throne of God, or by the earth, for it is his footstool, or by Jerusalem, for it is the city of the great King. And do not take an oath by your head, for you cannot make one hair white or black. Let what you say be simply "Yes" or "No"; anything more than this comes from evil.

You have heard that it was said, "An eye for an eye and a tooth for a tooth." But I say to you, Do not resist the one who is evil. But if anyone slaps you on the right cheek, turn to him the other also. And if anyone would sue you and take your tunic, let him have your cloak as well. And if anyone forces you to go one mile, go with him two miles. Give to the one who begs from you, and do not refuse the one who would borrow from you.

You have heard that it was said, "You shall love your neighbor and hate your enemy." But I say to you, Love your enemies and pray for those who persecute you, so that you may be sons of your Father who is in heaven. For he makes his sun rise on the evil and on the good, and sends rain on the just and on the unjust. For if you love those who love you, what reward do you have? Do not even the tax collectors do the same? And if you greet only your brothers, what more are you doing than others? Do not even the Gentiles do the same?

You therefore must be perfect, as your heavenly Father is perfect.

Beware of practicing your righteousness before other people in order to be seen by them, for then you will have no reward from your Father who is in heaven.

Thus, when you give to the needy, sound no trumpet before you, as the hypocrites do in the synagogues and in the streets, that they may be praised by others. Truly, I say to you, they have received their reward. But when you give to the needy, do not let your left hand know what your right hand is doing, so that your giving may be in secret. And your Father who sees in secret will reward you.

And when you pray, you must not be like the hypocrites. For they love to stand and pray in the synagogues and at the street corners, that they may be seen by others. Truly, I say to you, they have received their reward. But when you pray, go into your room and shut the door and pray to your Father who is in secret. And your Father who sees in secret will reward you.

And when you pray, do not heap up empty phrases as the Gentiles do, for they think that they will be heard for their many words. Do not be like them, for your Father knows what you need before you ask him.

Pray then like this:

Our Father in heaven,

hallowed be your name.

Your kingdom come,

your will be done,

on earth as it is in heaven.

Give us this day our daily bread,

and forgive us our debts,

as we also have forgiven our debtors.

And lead us not into temptation,

but deliver us from evil.

For if you forgive others their trespasses, your heavenly Father will also forgive you, but if you do not forgive others their trespasses, neither will your Father forgive your trespasses.

And when you fast, do not look gloomy like the hypocrites, for they disfigure their faces that their fasting may be seen by others. Truly, I say to you, they have received their reward. But when you fast, anoint your head and wash your face, that your fasting may not be seen by others but by your Father who is in secret. And your Father who sees in secret will reward you.

Do not lay up for yourselves treasures on earth, where moth and rust destroy and where thieves break in and steal, but lay up for yourselves treasures in heaven, where neither moth nor rust destroys and where thieves do not break in and steal. For where your treasure is, there your heart will be also.

The eye is the lamp of the body. So, if your eye is healthy, your whole body will be full of light, but if your eye is bad, your whole body will be full of darkness. If then the light in you is darkness, how great is the darkness!

No one can serve two masters, for either he will hate the one and love the other, or he will be devoted to the one and despise the other. You cannot serve God and money.

Therefore I tell you, do not be anxious about your life, what you will eat or what you will drink, nor about your body, what you will put on. Is not life more than food, and the body more than clothing? Look at the birds of the air: they neither sow nor reap nor gather into barns, and yet your heavenly Father feeds them. Are you not of more value than they? And which of you by being anxious can add a single hour to his span of life?

And why are you anxious about clothing? Consider the lilies of the field, how they grow: they neither toil nor spin, yet I tell you, even Solomon in all his glory was not arrayed like one of these. But if God so clothes the grass of the field, which today is alive and tomorrow is thrown into the oven, will he not much more clothe you, O you of little faith?

Therefore do not be anxious, saying, "What shall we eat?" or "What shall we drink?" or "What shall we wear?" For the Gentiles seek after all these things, and your heavenly Father knows that you need them all. But seek first the kingdom of God and his righteousness, and all these things will be added to you.

Therefore do not be anxious about tomorrow, for tomorrow will be anxious for itself. Sufficient for the day is its own trouble.

judge not, that you be not judged.

For with the judgment you pronounce you will be judged, and with the measure you use it will be measured to you.

Why do you see the speck that is in your brother's eye, but do not notice the log that is in your own eye?

Or how can you say to your brother, "Let me take the speck out of your eye," when there is the log in your own eye?

You hypocrite, first take the log out of your own eye, and then you will see clearly to take the speck out of your brother's eye.

Do not give dogs what is holy, and do not throw your pearls before pigs, lest they trample them underfoot and turn to attack you.

Ask, and it will be given to you; seek, and you will find; knock, and it will be opened to you.

For everyone who asks receives, and the one who seeks finds, and to the one who knocks it will be opened.

Or which one of you, if his son asks him for bread, will give him a stone?

Or if he asks for a fish, will give him a serpent?

If you then, who are evil, know how to give good gifts to your children, how much more will your Father who is in heaven give good things to those who ask him!

So whatever you wish that others would do to you, do also to them, for this is the Law and the Prophets.

Enter by the narrow gate.

For the gate is wide and the way is easy that leads to destruction, and those who enter by it are many.

For the gate is narrow and the way is hard that leads to life, and those who find it are few.

Beware of false prophets, who come to you in sheep's clothing but inwardly are ravenous wolves.

You will recognize them by their fruits.

Are grapes gathered from thornbushes, or figs from thistles?

So, every healthy tree bears good fruit, but the diseased tree bears bad fruit.

A healthy tree cannot bear bad fruit, nor can a diseased tree bear good fruit.

Every tree that does not bear good fruit is cut down and thrown into the fire.

Thus you will recognize them by their fruits.

Not everyone who says to me, "Lord, Lord," will enter the kingdom of heaven, but the one who does the will of my Father who is in heaven.

On that day many will say to me, "Lord, Lord, did we not prophesy in your name, and cast out demons in your name, and do many mighty works in your name?"

And then will I declare to them, "I never knew you; depart from me, you workers of lawlessness."

Everyone then who hears these words of mine and does them will be like a wise man who built his house on the rock.

And the rain fell, and the floods came, and the winds blew and beat on that house, but it did not fall, because it had been founded on the rock.

And everyone who hears these words of mine and does not do them will be like a foolish man who built his house on the sand.

And the rain fell, and the floods came, and the winds blew and beat against that house, and it fell, and great was the fall of it.

And when Jesus finished these sayings, the crowds were astonished at his teaching,

for he was teaching them as one who had authority, and not as their scribes.

Luke 6:31: "And as you wish that others would do to you, do so to them."

Matthew 5:21–48: "You have heard that it was said to those of old, 'You shall not murder; and whoever murders will be liable to judgment.' But I say to you that everyone who is angry with his brother[a] will be liable to judgment; whoever insults[b] his brother will be liable to the council; and whoever says, 'You fool!' will be liable to the hell[c] of fire. So if you are offering your gift at the altar and there remember that your brother has something against you, leave your gift there before the altar and go. First be reconciled to your brother, and then come and offer your gift. Come to terms quickly with your accuser while you are going with him to court, lest your accuser hand you over to the judge, and the judge to the guard, and you be put in prison. Truly, I say to you, you will never get out until you have paid the last penny. "You have heard that it was said, 'You shall not commit adultery.' But I say to you that everyone who looks at a woman with lustful intent has already committed adultery with her in his heart. If your right eye causes you to sin, tear it out and throw it away. For it is better that you lose one of your members than that your whole body be thrown into hell. And if your right hand causes you to sin, cut it off and throw it away. For it is better that you lose one of your members than that your whole body go into hell. "It was also said, 'Whoever divorces his wife, let him give her a certificate of divorce.' But I say to you that everyone who divorces his wife, except on the ground of sexual immorality, makes her commit adultery, and whoever marries a divorced woman commits adultery. "Again you have heard that it was said to those of old, 'You shall not swear falsely, but shall perform to the Lord what you have sworn.' But I say to you, Do not take an oath at all, either by heaven, for it is the throne of God, or by the earth, for it is his footstool, or by Jerusalem, for it is the city of the great King. And do not take an oath by your head, for you cannot make one hair white or black. Let what you say be simply 'Yes' or 'No'; anything more than this comes from evil. "You have heard that it was said, 'An eye for an eye and a tooth for a tooth.'

But I say to you, Do not resist the one who is evil. But if anyone slaps you on the right cheek, turn to him the other also. And if anyone would sue you and take your tunic,[f] let him have your cloak as well. And if anyone forces you to go one mile, go with him two miles. Give to the one who begs from you, and do not refuse the one who would borrow from you. "You have heard that it was said, 'You shall love your neighbor and hate your enemy.' But I say to you, Love your enemies and pray for those who persecute you, so that you may be sons of your Father who is in heaven. For he makes his sun rise on the evil and on the good, and sends rain on the just and on the unjust. For if you love those who love you, what reward do you have? Do not even the tax collectors do the same? And if you greet only your brothers,[g] what more are you doing than others? Do not even the Gentiles do the same? You therefore must be perfect, as your heavenly Father is perfect."

Matthew 21:12–13: "And Jesus entered the temple and drove out all who sold and bought in the temple, and he overturned the tables of the money-changers and the seats of those who sold pigeons. He said to them, 'It is written, "My house shall be called a house of prayer," but you make it a den of robbers.'"

Luke 19:10: "For the Son of Man came to seek and to save the lost."

John 13:1–17: Now before the Feast of the Passover, when Jesus knew that his hour had come to depart out of this world to the Father, having loved his own who were in the world, he loved them to the end. During supper, when the devil had already put it into the heart of Judas Iscariot, Simon's son, to betray him, Jesus, knowing that the Father had given all things into his hands, and that he had come from God and was going back to God, rose from supper. He laid aside his outer garments, and taking a towel, tied it around his waist. Then he poured water into a basin and began to wash the disciples' feet and to wipe them with the towel that was wrapped around him. He came to Simon Peter, who said to him, "Lord, do you wash my feet?" Jesus answered him, "What I am doing you do not understand now, but afterward you will understand." Peter said to him, "You shall never wash my feet." Jesus answered him, "If I do not wash you, you have no share with me." Simon Peter said to him, "Lord, not my feet only but also my hands and my head!" Jesus said to him, "The one who has bathed does not need to wash, except for his feet,[a] but is completely clean. And you[b] are clean, but not every one of you." For he knew who was to betray him; that was why he said, "Not all of you are clean."

When he had washed their feet and put on his outer garments and resumed his place, he said to them, "Do you understand what I have done to you? You call me Teacher and Lord, and you are right, for so I am. If I then, your Lord and Teacher, have washed your feet, you also ought to wash one another's feet. For I have given you an example, that you also should do just as I have done to you. Truly, truly, I say to you, a servant[c] is not greater than his master, nor is a messenger greater than the one who sent him. If you know these things, blessed are you if you do them."

Luke 23:34: "And Jesus said, 'Father, forgive them, for they know not what they do.' And they cast lots to divide his garments."

Mark 2:5–7: "And when Jesus saw their faith, he said to the paralytic, 'Son, your sins are forgiven.' Now some of the scribes were sitting there, questioning in their hearts, 'Why does this man speak like that? He is blaspheming! Who can forgive sins but God alone?'"

John 10:30: "I and the Father are one."

Matthew 14:33: "And those in the boat worshiped him, saying, 'Truly you are the Son of God.'"

Chapter 4
Jesus was a Legend

Acts 3:13–15: "The God of Abraham, the God of Isaac, and the God of Jacob, the God of our fathers, glorified his servant Jesus, whom you delivered over and denied in the presence of Pilate… you killed the Author of life, whom God raised from the dead. To this we are witnesses."

Matthew 28:5–7: "But the angel said to the women, "Do not be afraid, for I know that you seek Jesus who was crucified. He is not here, for he has risen, as he said. Come, see the place where he lay. Then go quickly and tell his disciples that he has risen from the dead, and behold, he is going before you to Galilee; there you will see him. See, I have told you.'"

Luke 6:27–28: "But I say to you who hear, Love your enemies, do good to those who hate you, bless those who curse you, pray for those who abuse you."

Mark 15:15: "So Pilate, wishing to satisfy the crowd, released for them Barabbas, and having scourged Jesus, he delivered him to be crucified."

2 Peter 1:16: "For we did not follow cleverly devised myths when we made known to you the power and coming of our Lord Jesus Christ, but we were eyewitnesses of his majesty."

Acts 5:40–42: "and when they had called in the apostles, they beat them and charged them not to speak in the name of Jesus, and let them go. Then they left the presence of the council, rejoicing that they were counted worthy to suffer dishonor for the name. And every day, in the temple and from house to house, they did not cease teaching and preaching that the Christ is Jesus."

Galatians 1:19: "But I saw none of the other apostles except James the Lord's brother."

Chapter 5
Was Jesus Just Mistaken?

Luke 22:42: "saying, 'Father, if you are willing, remove this cup from me. Nevertheless, not my will, but yours, be done.'"

John 8:58: "Jesus said to them, 'Truly, truly, I say to you, before Abraham was, I am.'"

Mark 2:5–7: "And when Jesus saw their faith, he said to the paralytic, 'Son, your sins are forgiven.' Now some of the scribes were sitting there, questioning in their hearts, 'Why does this man speak like that? He is blaspheming! Who can forgive sins but God alone?'"

John 14:9: "Jesus said to him, 'Have I been with you so long, and you still do not know me, Philip? Whoever has seen me has seen the Father. How can you say, "Show us the Father"?'"

Chapter 6
Jesus wasn't God

John 8:58: "Jesus said to them, 'Truly, truly, I say to you, before Abraham was, I am.'"

Exodus 3:14: "God said to Moses, 'I AM WHO I AM.' And he said, 'Say this to the people of Israel: I AM has sent me to you.'"

Mark 13:32: But concerning that day or that hour, no one knows, not even the angels in heaven, nor the Son, but only the Father.

Philippians 2:6-7: "who, though he was in the form of God, did not count equality with God a thing to be grasped, but emptied himself, by taking the form of a servant, being born in the likeness of men."
Daniel 7:13-14: "I saw in the night visions,and behold, with the clouds of heaven there came one like a son of man, and he came to the Ancient of Days and was presented before him. And to him was given dominion and glory and a kingdom, that all peoples, nations, and languages should serve him; his dominion is an everlasting dominion, which shall not pass away, and his kingdom one that shall not be destroyed."

Isaiah 9:6: For to us a child is born, to us a son is given; and the government shall be upon[a] his shoulder, and his name shall be called[b] Wonderful Counselor, Mighty God, Everlasting Father, Prince of Peace.

Psalm 110:1: "The Lord says to my Lord: "Sit at my right hand, until I make your enemies your footstool.""

John 4:25-26 "The woman said to him, "I know that Messiah is coming (he who is called Christ). When he comes, he will tell us all things." Jesus said to her, "I who speak to you am he."

John 8:58: Jesus said to them, "Truly, truly, I say to you, before Abraham was, I am."

John 10:30-33: "I and the Father are one. The Jews picked up stones again to stone him. Jesus answered them, 'I have shown you many good works from the Father; for which of them are you going to stone me?' The Jews answered him, 'It is not for a good work that we are going to stone you but for blasphemy, because you, being a man, make yourself God.'"

John 1:1: "In the beginning was the Word, and the Word was with God, and the Word was God."

John 1:14: "And the Word became flesh and dwelt among us, and we have seen his glory, glory as of the only Son from the Father, full of grace and truth."

Colossians 2:9: "For in him the whole fullness of deity dwells bodily,"

Titus 2:13: "waiting for our blessed hope, the appearing of the glory of our great God and Savior Jesus Christ,"

Hebrews 1:3: "He is the radiance of the glory of God and the exact imprint of his nature, and he upholds the universe by the word of his power. After making purification for sins, he sat down at the right hand of the Majesty on high,"

Philippians 2:6-7: who, though he was in the form of God, did not count equality with God a thing to be grasped, but emptied himself, by taking the form of a servant, being born in the likeness of men.

Revelation 1:8: "'I am the Alpha and the Omega,' says the Lord God, 'who is and who was and who is to come, the Almighty.'"

Revelation 22:13: "I am the Alpha and the Omega, the first and the last, the beginning and the end."

Isaiah 44:6: "Thus says the Lord, the King of Israel and his Redeemer, the Lord of hosts: 'I am the first and I am the last; besides me there is no god.'"

Deuteronomy 6:13: "It is the Lord your God you shall fear. Him you shall serve and by his name you shall swear."

Exodus 34:14: "for you shall worship no other god, for the Lord, whose name is Jealous, is a jealous God,"

Matthew 14:33: "And those in the boat worshiped him, saying, 'Truly you are the Son of God.'"

John 20:28: "Thomas answered him, 'My Lord and my God!'"

Matthew 28:9: "And behold, Jesus met them and said, 'Greetings!' And they came up and took hold of his feet and worshiped him."

Matthew 28:17: "And when they saw him they worshiped him, but some doubted."

Matthew 8:23-27: "And when he got into the boat, his disciples followed him. And behold, there arose a great storm on the sea, so that the boat was being swamped by the waves; but he was asleep. And they went and woke him, saying, 'Save us, Lord; we are perishing.' And he said to them, 'Why are you afraid, O you of little faith?' Then he rose and rebuked the winds and the sea, and there was a great calm. And the men marveled, saying, 'What sort of man is this, that even winds and sea obey him?'"

Mark 4:35-41: "On that day, when evening had come, he said to them, 'Let us go across to the other side.' And leaving the crowd, they took him with them in the boat, just as he was. And other boats were with him. And a great windstorm arose, and the waves were breaking into the boat, so that the boat was already filling. But he was in the stern, asleep on the cushion. And they woke him and said to him, 'Teacher, do you not care that we are perishing?' And he awoke and rebuked the wind and said to the sea, 'Peace! Be still!' And the wind ceased, and there was a great calm. He said to them, 'Why are you so afraid? Have you still no faith?' And they were filled with great fear and said to one another, 'Who then is this, that even the wind and the sea obey him?'"

118

Luke 8:22-25: "One day he got into a boat with his disciples, and he said to them, 'Let us go across to the other side of the lake.' So they set out, and as they sailed he fell asleep. And a windstorm came down on the lake, and they were filling with water and were in danger. And they went and woke him, saying, 'Master, Master, we are perishing!' And he awoke and rebuked the wind and the raging waves, and they ceased, and there was a calm. He said to them, 'Where is your faith?' And they were afraid, and they marveled, saying to one another, 'Who then is this, that he commands even winds and water, and they obey him?'"

Psalm 89:8-9: "O Lord God of hosts, who is mighty as you are, O Lord, with your faithfulness all around you? You rule the raging of the sea; when its waves rise, you still them."

Job 38:8-11: "Or who shut in the sea with doors when it burst out from the womb, when I made clouds its garment and thick darkness its swaddling band, and prescribed limits for it and set bars and doors, and said, 'Thus far shall you come, and no farther, and here shall your proud waves be stayed'?"

Matthew 14:13-21: "Now when Jesus heard this, he withdrew from there in a boat to a desolate place by himself. But when the crowds heard it, they followed him on foot from the towns. When he went ashore he saw a great crowd, and he had compassion on them and healed their sick. Now when it was evening, the disciples came to him and said, 'This is a desolate place, and the day is now over; send the crowds away to go into the villages and buy food for themselves.' But Jesus said, 'They need not go away; you give them something to eat.' They said to him, 'We have only five loaves here and two fish.' And he said, 'Bring them here to me.' Then he ordered the crowds to sit down on the grass, and taking the five loaves and the two fish, he looked up to heaven and said a blessing. Then he broke the loaves and gave them to the disciples, and the disciples gave them to the crowds. And they all ate and were satisfied. And they took up twelve baskets full of the broken pieces left over. And those who ate were about five thousand men, besides women and children."

John 6:1-14: "After this Jesus went away to the other side of the Sea of Galilee, which is the Sea of Tiberias. And a large crowd was following him, because they saw the signs that he was doing on the sick. Jesus went up on the mountain, and there he sat down with his disciples. Now the Passover, the feast of the Jews, was at hand. Lifting up his eyes, then, and seeing that a large crowd was coming toward him, Jesus said to Philip, 'Where are we to buy bread, so that these people may eat?' He said this to test him, for he himself knew what he would do. Philip answered him, 'Two hundred denarii worth of bread would not be enough for each of them to get a little.' One of his disciples, Andrew, Simon Peter's brother, said to him, 'There is a boy here who has five barley loaves and two fish, but what are they for so many?' Jesus said, 'Have the people sit down.' Now there was much grass in the place. So the men sat down, about five thousand in number. Jesus then took the loaves, and when he had given thanks, he distributed them to those who were seated. So also the fish, as much as they wanted. And when they had eaten their fill, he told his disciples, 'Gather up the leftover fragments, that nothing may be lost.' So they gathered them up and filled twelve baskets with fragments from the five barley loaves left by those who had eaten. When the people saw the sign that he had done, they said, 'This is indeed the Prophet who is to come into the world!'"

John 6:35: "Jesus said to them, 'I am the bread of life; whoever comes to me shall not hunger, and whoever believes in me shall never thirst.'"

Matthew 14:22-33: "Immediately he made the disciples get into the boat and go before him to the other side, while he dismissed the crowds. And after he had dismissed the crowds, he went up on the mountain by himself to pray. When evening came, he was there alone, but the boat by this time was a long way from the land, beaten by the waves, for the wind was against them. And in the fourth watch of the night he came to them, walking on the sea. But when the disciples saw him walking on the sea, they were terrified, and said, 'It is a ghost!' and they cried out in fear. But immediately Jesus spoke to them, saying, 'Take heart; it is I. Do not be afraid.' And Peter answered him, 'Lord, if it is you, command me to come to you on the water.' He said, 'Come.' So Peter got out of the boat and walked on the water and came to Jesus. But when he saw the wind, he was afraid, and beginning to sink he cried out, 'Lord, save me.' Jesus immediately reached out his hand and took hold of him, saying to him, 'O you of little faith, why did you doubt?' And when they got into the boat, the wind ceased. And those in the boat worshiped him, saying, 'Truly you are the Son of God.'"

Mark 6:45-52: "Immediately he made his disciples get into the boat and go before him to the other side, to Bethsaida, while he dismissed the crowd. And after he had taken leave of them, he went up on the mountain to pray. And when evening came, the boat was out on the sea, and he was alone on the land. And he saw that they were making headway painfully, for the wind was against them. And about the fourth watch of the night he came to them, walking on the sea. He meant to pass by them, but when they saw him walking on the sea they thought it was a ghost, and cried out, for they all saw him and were terrified. But immediately he spoke to them and said, 'Take heart; it is I. Do not be afraid.' And he got into the boat with them, and the wind ceased. And they were utterly astounded, for they did not understand about the loaves, but their hearts were hardened."

John 6:16-21: "When evening came, his disciples went down to the sea, got into a boat, and started across the sea to Capernaum. It was now dark, and Jesus had not yet come to them. The sea became rough because a strong wind was blowing. When they had rowed about three or four miles, they saw Jesus walking on the sea and coming near the boat, and they were frightened. But he said to them, 'It is I; do not be afraid.' Then they were glad to take him into the boat, and immediately the boat was at the land to which they were going."

John 9: "As he passed by, he saw a man blind from birth. And his disciples asked him, 'Rabbi, who sinned, this man or his parents, that he was born blind?' Jesus answered, 'It was not that this man sinned, or his parents, but that the works of God might be displayed in him. We must work the works of him who sent me while it is day; night is coming, when no one can work. As long as I am in the world, I am the light of the world.' Having said these things, he spit on the ground and made mud with the saliva. Then he anointed the man's eyes with the mud and said to him, 'Go, wash in the pool of Siloam' (which means Sent). So he went and washed and came back seeing.

The neighbors and those who had seen him before as a beggar were saying, 'Is this not the man who used to sit and beg?' Some said, 'It is he.' Others said, 'No, but he is like him.' He kept saying, 'I am the man.' So they said to him, 'Then how were your eyes opened?' He answered, 'The man called Jesus made mud and anointed my eyes and said to me, "Go to Siloam and wash." So I went and washed and received my sight.' They said to him, 'Where is he?' He said, 'I do not know.'

They brought to the Pharisees the man who had formerly been blind. Now it was a Sabbath day when Jesus made the mud and opened his eyes. So the Pharisees again asked him how he had received his sight. And he said to them, 'He put mud on my eyes, and I washed, and I see.' Some of the Pharisees said, 'This man is not from God, for he does not keep the Sabbath.' But others said, 'How can a man who is a sinner do such signs?' And there was a division among them. So they said again to the blind man, 'What do you say about him, since he has opened your eyes?' He said, 'He is a prophet.'

The Jews did not believe that he had been blind and had received his sight, until they called the parents of the man who had received his sight and asked them, 'Is this your son, who you say was born blind? How then does he now see?' His parents answered, 'We know that this is our son and that he was born blind. But how he now sees we do not know, nor do we know who opened his eyes. Ask him; he is of age. He will speak for himself.' (His parents said these things because they feared the Jews, for the Jews had already agreed that if anyone should confess Jesus to be Christ, he was to be put out of the synagogue.) Therefore his parents said, 'He is of age; ask him.'

So for the second time they called the man who had been blind and said to him, 'Give glory to God. We know that this man is a sinner.' He answered, 'Whether he is a sinner I do not know. One thing I do know, that though I was blind, now I see.' They said to him, 'What did he do to you? How did he open your eyes?' He answered them, 'I have told you already, and you would not listen. Why do you want to hear it again? Do you also want to become his disciples?' And they reviled him, saying, 'You are his disciple, but we are disciples of Moses. We know that God has spoken to Moses, but as for this man, we do not know where he comes from.' The man answered, 'Why, this is an amazing thing! You do not know where he comes from, and yet he opened my eyes. We know that God does not listen to sinners, but if anyone is a worshiper of God and does his will, God listens to him. Never since the world began has it been heard that anyone opened the eyes of a man born blind. If this man were not from God, he could do nothing.' They answered him, 'You were born in utter sin, and would you teach us?' And they cast him out.

Jesus heard that they had cast him out, and having found him he said, 'Do you believe in the Son of Man?' He answered, 'And who is he, sir, that I may believe in him?' Jesus said to him, 'You have seen him, and it is he who is speaking to you.' He said, 'Lord, I believe,' and he worshiped him. Jesus said, 'For judgment I came into this world, that those who do not see may see, and those who see may become blind.' Some of the Pharisees near him heard these things and said to him, 'Are we also blind?' Jesus said to them, 'If you were blind, you would have no guilt; but now that you say, "We see," your guilt remains.'"

Matthew 8:1-4: "When he came down from the mountain, great crowds followed him. And behold, a leper came to him and knelt before him, saying, 'Lord, if you will, you can make me clean.' And Jesus stretched out his hand and touched him, saying, 'I will; be clean.' And immediately his leprosy was cleansed. And Jesus said to him, 'See that you say nothing to anyone, but go, show yourself to the priest and offer the gift that Moses commanded, for a proof to them.'"

Mark 2:1-12: "And when he returned to Capernaum after some days, it was reported that he was at home. And many were gathered together, so that there was no more room, not even at the door. And he was preaching the word to them. And they came, bringing to him a paralytic carried by four men. And when they could not get near him because of the crowd, they removed the roof above him, and when they had made an opening, they let down the bed on which the paralytic lay. And when Jesus saw their faith, he said to the paralytic, 'Son, your sins are forgiven.' Now some of the scribes were sitting there, questioning in their hearts, 'Why does this man speak like that? He is blaspheming! Who can forgive sins but God alone?' And immediately Jesus, perceiving in his spirit that they thus questioned within themselves, said to them, 'Why do you question these things in your hearts? Which is easier, to say to the paralytic, "Your sins are forgiven," or to say, "Rise, take up your bed and walk"? But that you may know that the Son of Man has authority on earth to forgive sins'—he said to the paralytic—'I say to you, rise, pick up your bed, and go home.' And he rose and immediately picked up his bed and went out before them all, so that they were all amazed and glorified God, saying, 'We never saw anything like this!'"

Matthew 8:1-4: "When he came down from the mountain, great crowds followed him. And behold, a leper came to him and knelt before him, saying, 'Lord, if you will, you can make me clean.' And Jesus stretched out his hand and touched him, saying, 'I will; be clean.' And immediately his leprosy was cleansed. And Jesus said to him, 'See that you say nothing to anyone, but go, show yourself to the priest and offer the gift that Moses commanded, for a proof to them.'"

Isaiah 35:5-6: "Then the eyes of the blind shall be opened, and the ears of the deaf unstopped; then shall the lame man leap like a deer, and the tongue of the mute sing for joy. For waters break forth in the wilderness, and streams in the desert;"

John 4:46-54: "So he came again to Cana in Galilee, where he had made the water wine. And at Capernaum there was an official whose son was ill. When this man heard that Jesus had come from Judea to Galilee, he went to him and asked him to come down and heal his son, for he was at the point of death. So Jesus said to him, 'Unless you see signs and wonders you will not believe.' The official said to him, 'Sir, come down before my child dies.' Jesus said to him, 'Go; your son will live.' The man believed the word that Jesus spoke to him and went on his way. As he was going down, his servants met him and told him that his son was recovering. So he asked them the hour when he began to get better, and they said to him, 'Yesterday at the seventh hour the fever left him.' The father knew that was the hour when Jesus had said to him, 'Your son will live.' And he himself believed, and all his household. This was now the second sign that Jesus did when he had come from Judea to Galilee."

Matthew 8:5-13: "When he had entered Capernaum, a centurion came forward to him, appealing to him, 'Lord, my servant is lying paralyzed at home, suffering terribly.' And he said to him, 'I will come and heal him.' But the centurion replied, 'Lord, I am not worthy to have you come under my roof, but only say the word, and my servant will be healed. For I too am a man under authority, with soldiers under me. And I say to one, "Go," and he goes, and to another, "Come," and he comes, and to my servant, "Do this," and he does it.' When Jesus heard this, he marveled and said to those who followed him, 'Truly, I tell you, with no one in Israel have I found such faith. I tell you, many will come from east and west and recline at table with Abraham, Isaac, and Jacob in the kingdom of heaven, while the sons of the kingdom will be thrown into the outer darkness. In that place there will be weeping and gnashing of teeth.' And to the centurion Jesus said, 'Go; let it be done for you as you have believed.' And the servant was healed at that very moment."

Mark 1:21-28: "And they went into Capernaum, and immediately on the Sabbath he entered the synagogue and was teaching. And they were astonished at his teaching, for he taught them as one who had authority, and not as the scribes. And immediately there was in their synagogue a man with an unclean spirit. And he cried out, 'What have you to do with us, Jesus of Nazareth? Have you come to destroy us? I know who you are—the Holy One of God.' But Jesus rebuked him, saying, 'Be silent, and come out of him!' And the unclean spirit, convulsing him and crying out with a loud voice, came out of him. And they were all amazed, so that they questioned among themselves, saying, 'What is this? A new teaching with authority! He commands even the unclean spirits, and they obey him.' And at once his fame spread everywhere throughout all the surrounding region of Galilee."

Matthew 8:28-34: "And when he came to the other side, to the country of the Gadarenes, two demon-possessed men met him, coming out of the tombs, so fierce that no one could pass that way. And behold, they cried out, 'What have you to do with us, O Son of God? Have you come here to torment us before the time?' Now a herd of many pigs was feeding at some distance from them. And the demons begged him, saying, 'If you cast us out, send us away into the herd of pigs.' And he said to them, 'Go.' So they came out and went into the pigs, and behold, the whole herd rushed down the steep bank into the sea and drowned in the waters. The herdsmen fled, and going into the city they told everything, especially what had happened to the demon-possessed men. And behold, all the city came out to meet Jesus, and when they saw him, they begged him to leave their region."

Luke 11:14-22: "Now he was casting out a demon that was mute. When the demon had gone out, the mute man spoke, and the people marveled. But some of them said, 'He casts out demons by Beelzebul, the prince of demons,' while others, to test him, kept seeking from him a sign from heaven. But he, knowing their thoughts, said to them, 'Every kingdom divided against itself is laid waste, and a divided household falls. And if Satan also is divided against himself, how will his kingdom stand? For you say that I cast out demons by Beelzebul. And if I cast out demons by Beelzebul, by whom do your sons cast them out? Therefore they will be your judges. But if it is by the finger of God that I cast out demons, then the kingdom of God has come upon you. When a strong man, fully armed, guards his own palace, his goods are safe; but when one stronger than he attacks him and overcomes him, he takes away his armor in which he trusted and divides his spoil.'"

Mark 5:35-43: "While he was still speaking, there came from the ruler's house some who said, 'Your daughter is dead. Why trouble the Teacher any further?' But overhearing what they said, Jesus said to the ruler of the synagogue, 'Do not fear, only believe.' And he allowed no one to follow him except Peter and James and John the brother of James. They came to the house of the ruler of the synagogue, and Jesus saw a commotion, people weeping and wailing loudly. And when he had entered, he said to them, 'Why are you making a commotion and weeping? The child is not dead but sleeping.' And they laughed at him. But he put them all outside and took the child's father and mother and those who were with him and went in where the child was. Taking her by the hand he said to her, 'Talitha cumi,' which means, 'Little girl, I say to you, arise.' And immediately the girl got up and began walking (for she was twelve years of age), and they were immediately overcome with amazement. And he strictly charged them that no one should know this, and told them to give her something to eat."

Luke 8:40-56: "Now when Jesus returned, the crowd welcomed him, for they were all waiting for him. And there came a man named Jairus, who was a ruler of the synagogue. And falling at Jesus' feet, he implored him to come to his house, for he had an only daughter, about twelve years of age, and she was dying. As Jesus went, the people pressed around him. And there was a woman who had had a discharge of blood for twelve years, and though she had spent all her living on physicians, she could not be healed by anyone. She came up behind him and touched the fringe of his garment, and immediately her discharge of blood ceased. And Jesus said, 'Who was it that touched me?' When all denied it, Peter said, 'Master, the crowds surround you and are pressing in on you!' But Jesus said, 'Someone touched me, for I perceive that power has gone out from me.' And when the woman saw that she was not hidden, she came trembling, and falling down before him declared in the presence of all the people why she had touched him, and how she had been immediately healed. And he said to her, 'Daughter, your faith has made you well; go in peace.' While he was still speaking, someone from the ruler's house came and said, 'Your daughter is dead; do not trouble the Teacher anymore.' But Jesus on hearing this answered him, 'Do not fear; only believe, and she will be well.' And when he came to the house, he allowed no one to enter with him, except Peter and John and James, and the father and mother of the child. And all were weeping and mourning for her, but he said, 'Do not weep, for she is not dead but sleeping.' And they laughed at him, knowing that she was dead. But taking her by the hand he called, saying, 'Child, arise.' And her spirit returned, and she got up at once. And he directed that something should be given her to eat. And her parents were amazed, but he charged them to tell no one what had happened."

Luke 7:11-17: "Soon afterward he went to a town called Nain, and his disciples and a great crowd went with him. As he drew near to the gate of the town, behold, a man who had died was being carried out, the only son of his mother, and she was a widow, and a considerable crowd from the town was with her. And when the Lord saw her, he had compassion on her and said to her, 'Do not weep.' Then he came up and touched the bier, and the bearers stood still. And he said, 'Young man, I say to you, arise.' And the dead man sat up and began to speak, and Jesus gave him to his mother. Fear seized them all, and they glorified God, saying, 'A great prophet has arisen among us!' and 'God has visited his people!' And this report about him spread through the whole of Judea and all the surrounding country."

John 11:1-44: "Now a certain man was ill, Lazarus of Bethany, the village of Mary and her sister Martha. It was Mary who anointed the Lord with ointment and wiped his feet with her hair, whose brother Lazarus was ill. So the sisters sent to him, saying, 'Lord, he whom you love is ill.' But when Jesus heard it he said, 'This illness does not lead to death. It is for the glory of God, so that the Son of God may be glorified through it.'

Now Jesus loved Martha and her sister and Lazarus. So, when he heard that Lazarus was ill, he stayed two days longer in the place where he was. Then after this he said to the disciples, 'Let us go to Judea again.' The disciples said to him, 'Rabbi, the Jews were just now seeking to stone you, and are you going there again?' Jesus answered, 'Are there not twelve hours in the day? If anyone walks in the day, he does not stumble, because he sees the light of this world. But if anyone walks in the night, he stumbles, because the light is not in him.' After saying these things, he said to them, 'Our friend Lazarus has fallen asleep, but I go to awaken him.' The disciples said to him, 'Lord, if he has fallen asleep, he will recover.' Now Jesus had spoken of his death, but they thought that he meant taking rest in sleep. Then Jesus told them plainly, 'Lazarus has died, and for your sake I am glad that I was not there, so that you may believe. But let us go to him.' So Thomas, called the Twin, said to his fellow disciples, 'Let us also go, that we may die with him.'

Now when Jesus came, he found that Lazarus had already been in the tomb four days. Bethany was near Jerusalem, about two miles off, and many of the Jews had come to Martha and Mary to console them concerning their brother. So when Martha heard that Jesus was coming, she went and met him, but Mary remained seated in the house. Martha said to Jesus, 'Lord, if you had been here, my brother would not have died. But even now I know that whatever you ask from God, God will give you.' Jesus said to her, 'Your brother will rise again.' Martha said to him, 'I know that he will rise again in the resurrection on the last day.' Jesus said to her, 'I am the resurrection and the life. Whoever believes in me, though he die, yet shall he live, and everyone who lives and believes in me shall never die. Do you believe this?' She said to him, 'Yes, Lord; I believe that you are the Christ, the Son of God, who is coming into the world.'

When she had said this, she went and called her sister Mary, saying in private, 'The Teacher is here and is calling for you.' And when she heard it, she rose quickly and went to him. Now Jesus had not yet come into the village, but was still in the place where Martha had met him. When the Jews who were with her in the house, consoling her, saw Mary rise quickly and go out, they followed her, supposing that she was going to the tomb to weep there.

Now when Mary came to where Jesus was and saw him, she fell at his feet, saying to him, 'Lord, if you had been here, my brother would not have died.' When Jesus saw her weeping, and the Jews who had come with her also weeping, he was deeply moved in his spirit and greatly troubled. And he said, 'Where have you laid him?' They said to him, 'Lord, come and see.' Jesus wept. So the Jews said, 'See how he loved him!' But some of them said, 'Could not he who opened the eyes of the blind man also have kept this man from dying?'

Then Jesus, deeply moved again, came to the tomb. It was a cave, and a stone lay against it. Jesus said, 'Take away the stone.' Martha, the sister of the dead man, said to him, 'Lord, by this time there will be an odor, for he has been dead four days.' Jesus said to her, 'Did I not tell you that if you believed you would see the glory of God?' So they took away the stone. And Jesus lifted up his eyes and said, 'Father, I thank you that you have heard me. I knew that you always hear me, but I said this on account of the people standing around, that they may believe that you sent me.' When he had said these things, he cried out with a loud voice, 'Lazarus, come out.' The man who had died came out, his hands and feet bound with linen strips, and his face wrapped with a cloth. Jesus said to them, 'Unbind him, and let him go.'"

John 11:25: "Jesus said to her, 'I am the resurrection and the life. Whoever believes in me, though he die, yet shall he live,'"

Mark 2:1-12: Earlier in this Chapter

Matthew 9:1-8: "And getting into a boat he crossed over and came to his own city. And behold, some people brought to him a paralytic, lying on a bed. And when Jesus saw their faith, he said to the paralytic, 'Take heart, my son; your sins are forgiven.' And behold, some of the scribes said to themselves, 'This man is blaspheming.' But Jesus, knowing their thoughts, said, 'Why do you think evil in your hearts? For which is easier, to say, "Your sins are forgiven," or to say, "Rise and walk"? But that you may know that the Son of Man has authority on earth to forgive sins'—he then said to the paralytic—'Rise, pick up your bed and go home.' And he rose and went home. When the crowds saw it, they were afraid, and they glorified God, who had given such authority to men."

Luke 5:17-26: "On one of those days, as he was teaching, Pharisees and teachers of the law were sitting there, who had come from every village of Galilee and Judea and from Jerusalem. And the power of the Lord was with him to heal. And behold, some men were bringing on a bed a man who was paralyzed, and they were seeking to bring him in and lay him before Jesus, but finding no way to bring him in, because of the crowd, they went up on the roof and let him down with his bed through the tiles into the midst before Jesus. And when he saw their faith, he said, 'Man, your sins are forgiven you.' And the scribes and the Pharisees began to question, saying, 'Who is this who speaks blasphemies? Who can forgive sins but God alone?' When Jesus perceived their thoughts, he answered them, 'Why do you question in your hearts? Which is easier, to say, "Your sins are forgiven you," or to say, "Rise and walk"? But that you may know that the Son of Man has authority on earth to forgive sins'—he said to the man who was paralyzed—'I say to you, rise, pick up your bed and go home.' And immediately he rose up before them and picked up what he had been lying on and went home, glorifying God. And amazement seized them all, and they glorified God and were filled with awe, saying, 'We have seen extraordinary things today.'"

Luke 7:36-50: "One of the Pharisees asked him to eat with him, and he went into the Pharisee's house and reclined at table. And behold, a woman of the city, who was a sinner, when she learned that he was reclining at table in the Pharisee's house, brought an alabaster flask of ointment, and standing behind him at his feet, weeping, she began to wet his feet with her tears and wiped them with the hair of her head and kissed his feet and anointed them with the ointment. Now when the Pharisee who had invited him saw this, he said to himself, 'If this man were a prophet, he would have known who and what sort of woman this is who is touching him, for she is a sinner.' And Jesus answering said to him, 'Simon, I have something to say to you.' And he answered, 'Say it, Teacher.'

'A certain moneylender had two debtors. One owed five hundred denarii, and the other fifty. When they could not pay, he cancelled the debt of both. Now which of them will love him more?' Simon answered, 'The one, I suppose, for whom he cancelled the larger debt.' And he said to him, 'You have judged rightly.' Then turning toward the woman he said to Simon, 'Do you see this woman? I entered your house; you gave me no water for my feet, but she has wet my feet with her tears and wiped them with her hair. You gave me no kiss, but from the time I came in she has not ceased to kiss my feet. You did not anoint my head with oil, but she has anointed my feet with ointment. Therefore I tell you, her sins, which are many, are forgiven—for she loved much. But he who is forgiven little, loves little.' And he said to her, 'Your sins are forgiven.' Then those who were at table with him began to say among themselves, 'Who is this, who even forgives sins?' And he said to the woman, 'Your faith has saved you; go in peace.'"

1 Corinthians 15:3-8: "For I delivered to you as of first importance what I also received: that Christ died for our sins in accordance with the Scriptures, that he was buried, that he was raised on the third day in accordance with the Scriptures, and that he appeared to Cephas, then to the twelve. Then he appeared to more than five hundred brothers at one time, most of whom are still alive, though some have fallen asleep. Then he appeared to James, then to all the apostles. Last of all, as to one untimely born, he appeared also to me."

Acts 2:32: "This Jesus God raised up, and of that we all are witnesses."

Romans 1:4: "and was declared to be the Son of God in power according to the Spirit of holiness by his resurrection from the dead, Jesus Christ our Lord,"

John 2:19-21: "Jesus answered them, 'Destroy this temple, and in three days I will raise it up.' The Jews then said, 'It has taken forty-six years to build this temple, and will you raise it up in three days?' But he was speaking about the temple of his body."

Matthew 28:5-7: "But the angel said to the women, 'Do not be afraid, for I know that you seek Jesus who was crucified. He is not here, for he has risen, as he said. Come, see the place where he lay. Then go quickly and tell his disciples that he has risen from the dead, and behold, he is going before you to Galilee; there you will see him. See, I have told you.'"

Luke 24:36-43: "As they were talking about these things, Jesus himself stood among them, and said to them, 'Peace to you!' But they were startled and frightened and thought they saw a spirit. And he said to them, 'Why are you troubled, and why do doubts arise in your hearts? See my hands and my feet, that it is I myself. Touch me, and see. For a spirit does not have flesh and bones as you see that I have.' And when he had said this, he showed them his hands and his feet. And while they still disbelieved for joy and were marveling, he said to them, 'Have you anything here to eat?' They gave him a piece of broiled fish, and he took it and ate before them."

John 20:26-29: "Eight days later, his disciples were inside again, and Thomas was with them. Although the doors were locked, Jesus came and stood among them and said, 'Peace be with you.' Then he said to Thomas, 'Put your finger here, and see my hands; and put out your hand, and place it in my side. Do not disbelieve, but believe.' Thomas answered him, 'My Lord and my God!' Jesus said to him, 'Have you believed because you have seen me? Blessed are those who have not seen and yet have believed.'"

Chapter 7
Jesus was Mentally Ill

John 1:14: "And the Word became flesh and dwelt among us, and we have seen his glory, glory as of the only Son from the Father, full of grace and truth."

Matthew 5–7: See Chapter 3

Matthew 26:36-46: "Then Jesus went with them to a place called Gethsemane, and he said to his disciples, 'Sit here, while I go over there and pray.' And taking with him Peter and the two sons of Zebedee, he began to be sorrowful and troubled. Then he said to them, 'My soul is very sorrowful, even to death; remain here, and watch with me.' And going a little farther, he fell on his face and prayed, saying, 'My Father, if it be possible, let this cup pass from me; nevertheless, not as I will, but as you will.' And he came to the disciples and found them sleeping. And he said to Peter, 'So, could you not watch with me one hour? Watch and pray that you may not enter into temptation. The spirit indeed is willing, but the flesh is weak.' Again, for the second time, he went away and prayed, 'My Father, if this cannot pass unless I drink it, your will be done.' And again he came and found them sleeping, for their eyes were heavy. So, leaving them again, he went away and prayed for the third time, saying the same words again. Then he came to the disciples and said to them, 'Sleep and take your rest later on. See, the hour is at hand, and the Son of Man is betrayed into the hands of sinners. Rise, let us be going; see, my betrayer is at hand.'"

Mark 14:32-42: "And they went to a place called Gethsemane. And he said to his disciples, 'Sit here while I pray.' And he took with him Peter and James and John, and began to be greatly distressed and troubled. And he said to them, 'My soul is very sorrowful, even to death. Remain here and watch.' And going a little farther, he fell on the ground and prayed that, if it were possible, the hour might pass from him. And he said, 'Abba, Father, all things are possible for you. Remove this cup from me. Yet not what I will, but what you will.' And he came and found them sleeping, and he said to Peter, 'Simon, are you asleep? Could you not watch one hour? Watch and pray that you may not enter into temptation. The spirit indeed is willing, but the flesh is weak.' And again he went away and prayed, saying the same words. And again he came and found them sleeping, for their eyes were very heavy, and they did not know what to answer him. And he came the third time and said to them, 'Are you still sleeping and taking your rest? It is enough; the hour has come. The Son of Man is betrayed into the hands of sinners. Rise, let us be going; see, my betrayer is at hand.'"

Luke 22:39-46: "And he came out and went, as was his custom, to the Mount of Olives, and the disciples followed him. And when he came to the place, he said to them, 'Pray that you may not enter into temptation.' And he withdrew from them about a stone's throw, and knelt down and prayed, saying, 'Father, if you are willing, remove this cup from me. Nevertheless, not my will, but yours, be done.' And there appeared to him an angel from heaven, strengthening him. And being in agony he prayed more earnestly; and his sweat became like great drops of blood falling down to the ground. And when he rose from prayer, he came to the disciples and found them sleeping for sorrow, and he said to them, 'Why are you sleeping? Rise and pray that you may not enter into temptation.'"

John 18:1: "When Jesus had spoken these words, he went out with his disciples across the brook Kidron, where there was a garden, which he and his disciples entered."

Chapter 8
Jesus Didn't Say What We Think He Said

Mark 4:30-32: "And he said, 'With what can we compare the kingdom of God, or what parable shall we use for it? It is like a grain of mustard seed, which, when sown on the ground, is the smallest of all the seeds on earth, yet when it is sown it grows up and becomes larger than all the garden plants and puts out large branches, so that the birds of the air can make nests in its shade.'"

Matthew 13:31-32: "He put another parable before them, saying, 'The kingdom of heaven is like a grain of mustard seed that a man took and sowed in his field. It is the smallest of all seeds, but when it has grown it is larger than all the garden plants and becomes a tree, so that the birds of the air come and make nests in its branches.'"

Luke 13:18-19: "He said therefore, 'What is the kingdom of God like? And to what shall I compare it? It is like a grain of mustard seed that a man took and sowed in his garden, and it grew and became a tree, and the birds of the air made nests in its branches.'"

Matthew 5:3-12: "'Blessed are the poor in spirit, for theirs is the kingdom of heaven. Blessed are those who mourn, for they shall be comforted. Blessed are the meek, for they shall inherit the earth. Blessed are those who hunger and thirst for righteousness, for they shall be satisfied. Blessed are the merciful, for they shall receive mercy. Blessed are the pure in heart, for they shall see God. Blessed are the peacemakers, for they shall be called sons of God. Blessed are those who are persecuted for righteousness' sake, for theirs is the kingdom of heaven. Blessed are you when others revile you and persecute you and utter all kinds of evil against you falsely on my account. Rejoice and be glad, for your reward is great in heaven, for so they persecuted the prophets who were before you.'"

Luke 6:20-23: "And he lifted up his eyes on his disciples, and said: 'Blessed are you who are poor, for yours is the kingdom of God. Blessed are you who are hungry now, for you shall be satisfied. Blessed are you who weep now, for you shall laugh. Blessed are you when people hate you and when they exclude you and revile you and spurn your name as evil, on account of the Son of Man! Rejoice in that day, and leap for joy, for behold, your reward is great in heaven; for so their fathers did to the prophets.'"

Mark 14:36: "And he said, 'Abba, Father, all things are possible for you. Remove this cup from me. Yet not what I will, but what you will.'"

Mark 10:2-12: "And Pharisees came up and in order to test him asked, 'Is it lawful for a man to divorce his wife?' He answered them, 'What did Moses command you?' They said, 'Moses allowed a man to write a certificate of divorce and to send her away.' And Jesus said to them, 'Because of your hardness of heart he wrote you this commandment. But from the beginning of creation, "God made them male and female." "Therefore a man shall leave his father and mother and hold fast to his wife, and the two shall become one flesh." So they are no longer two but one flesh. What therefore God has joined together, let not man separate.' And in the house the disciples asked him again about this matter. And he said to them, 'Whoever divorces his wife and marries another commits adultery against her, and if she divorces her husband and marries another, she commits adultery.'"

Matthew 5:43-48: "'You have heard that it was said, "You shall love your neighbor and hate your enemy." But I say to you, Love your enemies and pray for those who persecute you, so that you may be sons of your Father who is in heaven. For he makes his sun rise on the evil and on the good, and sends rain on the just and on the unjust. For if you love those who love you, what reward do you have? Do not even the tax collectors do the same? And if you greet only your brothers, what more are you doing than others? Do not even the Gentiles do the same? You therefore must be perfect, as your heavenly Father is perfect.'"

Luke 6:27-36: "'But I say to you who hear, Love your enemies, do good to those who hate you, bless those who curse you, pray for those who abuse you. To one who strikes you on the cheek, offer the other also, and from one who takes away your cloak do not withhold your tunic either. Give to everyone who begs from you, and from one who takes away your goods do not demand them back. And as you wish that others would do to you, do so to them. If you love those who love you, what benefit is that to you? For even sinners love those who love them. And if you do good to those who do good to you, what benefit is that to you? For even sinners do the same. And if you lend to those from whom you expect to receive, what credit is that to you? Even sinners lend to sinners, to get back the same amount. But love your enemies, and do good, and lend, expecting nothing in return, and your reward will be great, and you

135

will be sons of the Most High, for he is kind to the ungrateful and the evil. Be merciful, even as your Father is merciful.'"

Mark 1:9-11: "In those days Jesus came from Nazareth of Galilee and was baptized by John in the Jordan. And when he came up out of the water, immediately he saw the heavens being torn open and the Spirit descending on him like a dove. And a voice came from heaven, 'You are my beloved Son; with you I am well pleased.'"

Matthew 3:13-17: "Then Jesus came from Galilee to the Jordan to John, to be baptized by him. John would have prevented him, saying, 'I need to be baptized by you, and do you come to me?' But Jesus answered him, 'Let it be so now, for thus it is fitting for us to fulfill all righteousness.' Then he consented. And when Jesus was baptized, immediately he went up from the water, and behold, the heavens were opened to him, and he saw the Spirit of God descending like a dove and coming to rest on him; and behold, a voice from heaven said, "This is my beloved Son,[b] with whom I am well pleased."

Luke 3:21-22: "Now when all the people were baptized, and when Jesus also had been baptized and was praying, the heavens were opened, and the Holy Spirit descended on him in bodily form, like a dove; and a voice came from heaven, 'You are my beloved Son; with you I am well pleased.'"

Mark 15:34: "And at the ninth hour Jesus cried with a loud voice, 'Eloi, Eloi, lema sabachthani?' which means, 'My God, my God, why have you forsaken me?'"

Mark 13:32: "But concerning that day or that hour, no one knows, not even the angels in heaven, nor the Son, but only the Father."

Chapter 9

Jesus isn't Lord

Romans 10:9: "because, if you confess with your mouth that Jesus is Lord and believe in your heart that God raised him from the dead, you will be saved."

Isaiah 53:3-5: "He was despised and rejected by men; a man of sorrows, and acquainted with grief; and as one from whom men hide their faces he was despised, and we esteemed him not. Surely he has borne our griefs and carried our sorrows; yet we esteemed him stricken, smitten by God, and afflicted. But he was pierced for our transgressions; he was crushed for our iniquities; upon him was the chastisement that brought us peace, and with his wounds we are healed."

2 Samuel 7:12-16: "When your days are fulfilled and you lie down with your fathers, I will raise up your offspring after you, who shall come from your body, and I will establish his kingdom. He shall build a house for my name, and I will establish the throne of his kingdom forever. I will be to him a father, and he shall be to me a son. When he commits iniquity, I will discipline him with the rod of men, with the stripes of the sons of men, but my steadfast love will not depart from him, as I took it from Saul, whom I put away from before you. And your house and your kingdom shall be made sure forever before me. Your throne shall be established forever.'"

John 20:24-29: "Now Thomas, one of the twelve, called the Twin, was not with them when Jesus came. So the other disciples told him, 'We have seen the Lord.' But he said to them, 'Unless I see in his hands the mark of the nails, and place my finger into the mark of the nails, and place my hand into his side, I will never believe.' Eight days later, his disciples were inside again, and Thomas was with them. Although the doors were locked, Jesus came and stood among them and said, 'Peace be with you.' Then he said to Thomas, 'Put your finger here, and see my hands; and put out your hand, and place it in my side. Do not disbelieve, but believe.' Thomas answered him, 'My Lord and my God!' Jesus said to him, 'Have you believed because you have seen me? Blessed are those who have not seen and yet have believed.'"

Mark 3:21: "And when his family heard it, they went out to seize him, for they were saying, 'He is out of his mind.'"

Acts 15:13-21: "After they finished speaking, James replied, 'Brothers, listen to me. Simeon has related how God first visited the Gentiles, to take from them a people for his name. And with this the words of the prophets agree, just as it is written, "After this I will return, and I will rebuild the tent of David that has fallen; I will rebuild its ruins, and I will restore it, that the remnant of mankind may seek the Lord, and all the Gentiles who are called by my name,

says the Lord, who makes these things known from of old."
Therefore my judgment is that we should not trouble those of the
Gentiles who turn to God, but should write to them to abstain from
the things polluted by idols, and from sexual immorality, and from
what has been strangled, and from blood. For from ancient
generations Moses has had in every city those who proclaim him,
for he is read every Sabbath in the synagogues.'"

Acts 9:1-19: "But Saul, still breathing threats and murder against
the disciples of the Lord, went to the high priest and asked him for
letters to the synagogues at Damascus, so that if he found any
belonging to the Way, men or women, he might bring them bound
to Jerusalem. Now as he went on his way, he approached Damascus,
and suddenly a light from heaven shone around him. And falling to
the ground, he heard a voice saying to him, 'Saul, Saul, why are you
persecuting me?' And he said, 'Who are you, Lord?' And he said, 'I
am Jesus, whom you are persecuting. But rise and enter the city, and
you will be told what you are to do.' The men who were traveling
with him stood speechless, hearing the voice but seeing no one. Saul
rose from the ground, and although his eyes were opened, he saw
nothing. So they led him by the hand and brought him into
Damascus. And for three days he was without sight, and neither ate
nor drank. Now there was a disciple at Damascus named Ananias.
The Lord said to him in a vision, 'Ananias.' And he said, 'Here I am,
Lord.' And the Lord said to him, 'Rise and go to the street called
Straight, and at the house of Judas look for a man of Tarsus named
Saul, for behold, he is praying, and he has seen in a vision a man
named Ananias come in and lay his hands on him so that he might
regain his sight.' But Ananias answered, 'Lord, I have heard from
many about this man, how much evil he has done to your saints at
Jerusalem. And here he has authority from the chief priests to bind
all who call on your name.' But the Lord said to him, 'Go, for he is
a chosen instrument of mine to carry my name before the Gentiles
and kings and the children of Israel. For I will show him how much
he must suffer for the sake of my name.' So Ananias departed and
entered the house. And laying his hands on him he said, 'Brother
Saul, the Lord Jesus who appeared to you on the road by which you
came has sent me so that you may regain your sight and be filled
with the Holy Spirit.' And immediately something like scales fell
from his eyes, and he regained his sight. Then he rose and was
baptized; and taking food, he was strengthened."

Acts 12:2: "He killed James the brother of John with the sword."

John 8:58: "Jesus said to them, 'Truly, truly, I say to you, before Abraham was, I am.'"

Mark 2:5-7: And when Jesus saw their faith, he said to the paralytic, "Son, your sins are forgiven." Now some of the scribes were sitting there, questioning in their hearts, "Why does this man speak like that? He is blaspheming! Who can forgive sins but God alone?"

Romans 10:9-10: because, if you confess with your mouth that Jesus is Lord and believe in your heart that God raised him from the dead, you will be saved. For with the heart one believes and is justified, and with the mouth one confesses and is saved.

Matthew 28:18: "And Jesus came and said to them, 'All authority in heaven and on earth has been given to me.'"

Acts 10:36: "As for the word that he sent to Israel, preaching good news of peace through Jesus Christ (he is Lord of all)."

John 14:6: "Jesus said to him, 'I am the way, and the truth, and the life. No one comes to the Father except through me.'

Chapter 10
Jesus didn't rise from the dead

1 Corinthians 15:3-8: For I delivered to you as of first importance what I also received: that Christ died for our sins in accordance with the Scriptures, that he was buried, that he was raised on the third day in accordance with the Scriptures, and that he appeared to Cephas, then to the twelve. Then he appeared to more than five hundred brothers at one time, most of whom are still alive, though some have fallen asleep. Then he appeared to James, then to all the apostles. Last of all, as to one untimely born, he appeared also to me.

Matthew 28:1-10: "Now after the Sabbath, toward the dawn of the first day of the week, Mary Magdalene and the other Mary went to see the tomb. And behold, there was a great earthquake, for an angel of the Lord descended from heaven and came and rolled back the stone and sat on it. His appearance was like lightning, and his clothing white as snow. And for fear of him the guards trembled and became like dead men. But the angel said to the women, 'Do not be afraid, for I know that you seek Jesus who was crucified. He is not here, for he has risen, as he said. Come, see the place where he lay. Then go quickly and tell his disciples that he has risen from the dead, and behold, he is going before you to Galilee; there you will see him. See, I have told you.' So they departed quickly from the tomb with fear and great joy, and ran to tell his disciples. And behold, Jesus met them and said, 'Greetings!' And they came up and took hold of his feet and worshiped him. Then Jesus said to them, 'Do not be afraid; go and tell my brothers to go to Galilee, and there they will see me.'"

Mark 16:1-8: "When the Sabbath was past, Mary Magdalene, Mary the mother of James, and Salome bought spices, so that they might go and anoint him. And very early on the first day of the week, when the sun had risen, they went to the tomb. And they were saying to one another, 'Who will roll away the stone for us from the entrance of the tomb?' And looking up, they saw that the stone had been rolled back—it was very large. And entering the tomb, they saw a young man sitting on the right side, dressed in a white robe, and they were alarmed. And he said to them, 'Do not be alarmed. You seek Jesus of Nazareth, who was crucified. He has risen; he is not here. See the place where they laid him. But go, tell his disciples and Peter that he is going before you to Galilee. There you will see him, just as he told you.' And they went out and fled from the tomb, for trembling and astonishment had seized them, and they said nothing to anyone, for they were afraid."

Luke 24:1-12: "But on the first day of the week, at early dawn, they went to the tomb, taking the spices they had prepared. And they found the stone rolled away from the tomb, but when they went in they did not find the body of the Lord Jesus. While they were perplexed about this, behold, two men stood by them in dazzling apparel. And as they were frightened and bowed their faces to the ground, the men said to them, 'Why do you seek the living among the dead? He is not here, but has risen. Remember how he told you, while he was still in Galilee, that the Son of Man must be delivered into the hands of sinful men and be crucified and on the third day rise.' And they remembered his words, and returning from the tomb they told all these things to the eleven and to all the rest. Now it was Mary Magdalene and Joanna and Mary the mother of James and the other women with them who told these things to the apostles, but these words seemed to them an idle tale, and they did not believe them. But Peter rose and ran to the tomb; stooping and looking in, he saw the linen cloths by themselves; and he went home marveling at what had happened."

John 20:1-10: "Now on the first day of the week Mary Magdalene came to the tomb early, while it was still dark, and saw that the stone had been taken away from the tomb. So she ran and went to Simon Peter and the other disciple, the one whom Jesus loved, and said to them, 'They have taken the Lord out of the tomb, and we do not know where they have laid him.' So Peter went out with the other disciple, and they were going toward the tomb. Both of them were running together, but the other disciple outran Peter and reached the tomb first. And stooping to look in, he saw the linen cloths lying there, but he did not go in. Then Simon Peter came, following him, and went into the tomb. He saw the linen cloths lying there, and the face cloth, which had been on Jesus' head, not lying with the linen cloths but folded up in a place by itself. Then the other disciple, who had reached the tomb first, also went in, and he saw and believed; for as yet they did not understand the Scripture, that he must rise from the dead. Then the disciples went back to their homes."

1 Corinthians 15:3-8: "For I delivered to you as of first importance what I also received: that Christ died for our sins in accordance with the Scriptures, that he was buried, that he was raised on the third day in accordance with the Scriptures, and that he appeared to Cephas, then to the twelve. Then he appeared to more than five hundred brothers at one time, most of whom are still alive, though some have fallen asleep. Then he appeared to James, then to all the apostles. Last of all, as to one untimely born, he appeared also to me."

John 20:27: Then he said to Thomas, "Put your finger here, and see my hands; and put out your hand, and place it in my side. Do not disbelieve, but believe."

Matthew 27:57-60: When it was evening, there came a rich man from Arimathea, named Joseph, who also was a disciple of Jesus. He went to Pilate and asked for the body of Jesus. Then Pilate ordered it to be given to him. And Joseph took the body and wrapped it in a clean linen shroud and laid it in his own new tomb, which he had cut in the rock. And he rolled a great stone to the entrance of the tomb and went away.

Matthew 28:11-15: While they were going, behold, some of the guard went into the city and told the chief priests all that had taken place. And when they had assembled with the elders and taken counsel, they gave a sufficient sum of money to the soldiers and said, "Tell people, 'His disciples came by night and stole him away while we were asleep.' And if this comes to the governor's ears, we will satisfy him and keep you out of trouble." So they took the money and did as they were directed. And this story has been spread among the Jews to this day.

Matthew 28:1-10: Now after the Sabbath, toward the dawn of the first day of the week, Mary Magdalene and the other Mary went to see the tomb. And behold, there was a great earthquake, for an angel of the Lord descended from heaven and came and rolled back the stone and sat on it. His appearance was like lightning, and his clothing white as snow. And for fear of him the guards trembled and became like dead men. But the angel said to the women, "Do not be afraid, for I know that you seek Jesus who was crucified. He is not here, for he has risen, as he said. Come, see the place where he lay. Then go quickly and tell his disciples that he has risen from the dead, and behold, he is going before you to Galilee; there you will see him. See, I have told you." So they departed quickly from the tomb with fear and great joy, and ran to tell his disciples. And behold, Jesus met them and said, "Greetings!" And they came up and took hold of his feet and worshiped him. Then Jesus said to them, "Do not be afraid; go and tell my brothers to go to Galilee, and there they will see me."

Luke 24:1-12: But on the first day of the week, at early dawn, they went to the tomb, taking the spices they had prepared. And they found the stone rolled away from the tomb, but when they went in they did not find the body of the Lord Jesus. While they were perplexed about this, behold, two men stood by them in dazzling apparel. And as they were frightened and bowed their faces to the ground, the men said to them, "Why do you seek the living among the dead? He is not here, but has risen. Remember how he told you, while he was still in Galilee, that the Son of Man must be delivered into the hands of sinful men and be crucified and on the third day rise." And they remembered his words, and returning from the tomb they told all these things to the eleven and to all the rest. Now it was Mary Magdalene and Joanna and Mary the mother of James and the other women with them who told these things to the apostles, but these words seemed to them an idle tale, and they did not believe them. But Peter rose and ran to the tomb; stooping and looking in, he saw the linen cloths by themselves; and he went home marveling at what had happened.

Acts 1:3: "He presented himself alive to them after his suffering by many proofs, appearing to them during forty days and speaking about the kingdom of God."

Romans 6:4: "We were buried therefore with him by baptism into death, in order that, just as Christ was raised from the dead by the glory of the Father, we too might walk in newness of life."

1 Thessalonians 4:14: "For since we believe that Jesus died and rose again, even so, through Jesus, God will bring with him those who have fallen asleep."

1 Peter 1:3: "Blessed be the God and Father of our Lord Jesus Christ! According to his great mercy, he has caused us to be born again to a living hope through the resurrection of Jesus Christ from the dead."

Revelation 1:17-18: "When I saw him, I fell at his feet as though dead. But he laid his right hand on me, saying, 'Fear not, I am the first and the last, and the living one. I died, and behold I am alive forevermore, and I have the keys of Death and Hades.'"

Matthew 27:51-53: "And behold, the curtain of the temple was torn in two, from top to bottom. And the earth shook, and the rocks were split. The tombs also were opened. And many bodies of the saints who had fallen asleep were raised, and coming out of the tombs after his resurrection they went into the holy city and appeared to many."

Matthew 28:5-6: "But the angel said to the women, 'Do not be afraid, for I know that you seek Jesus who was crucified. He is not here, for he has risen, as he said. Come, see the place where he lay.'"

Luke 24:6-7: "He is not here, but has risen. Remember how he told you, while he was still in Galilee, that the Son of Man must be delivered into the hands of sinful men and be crucified and on the third day rise."

John 11:25-26: "Jesus said to her, 'I am the resurrection and the life. Whoever believes in me, though he die, yet shall he live, and everyone who lives and believes in me shall never die. Do you believe this?'"

1 Corinthians 15:3-8: "For I delivered to you as of first importance what I also received: that Christ died for our sins in accordance with the Scriptures, that he was buried, that he was raised on the third day in accordance with the Scriptures, and that he appeared to Cephas, then to the twelve. Then he appeared to more than five hundred brothers at one time, most of whom are still alive, though some have fallen asleep. Then he appeared to James, then to all the apostles. Last of all, as to one untimely born, he appeared also to me."

1 Corinthians 15:20: "But in fact Christ has been raised from the dead, the firstfruits of those who have fallen asleep."

Romans 8:34: "Who is to condemn? Christ Jesus is the one who died—more than that, who was raised—who is at the right hand of God, who indeed is interceding for us."

Romans 14:9: "For to this end Christ died and lived again, that he might be Lord both of the dead and of the living."

Philippians 3:10: "That I may know him and the power of his resurrection, and may share his sufferings, becoming like him in his death."

Revelation 1:17b-18a: "'Fear not, I am the first and the last, and the living one. I died, and behold I am alive forevermore.'"

Mark 16:6: "And he said to them, 'Do not be alarmed. You seek Jesus of Nazareth, who was crucified. He has risen; he is not here. See the place where they laid him.'"

Matthew 28:7: "Then go quickly and tell his disciples that he has risen from the dead, and behold, he is going before you to Galilee; there you will see him. See, I have told you."

Matthew 27:52-53: "The tombs also were opened. And many bodies of the saints who had fallen asleep were raised, and coming out of the tombs after his resurrection they went into the holy city and appeared to many."

John 20:27: "Then he said to Thomas, 'Put your finger here, and see my hands; and put out your hand, and place it in my side. Do not disbelieve, but believe.'"

Acts 9:1-19: See Chapter 9

Chapter 11
Now You Have to Decide

John 14:6: "I am the way, and the truth, and the life. No one comes to the Father except through me."

Matthew 16:26: "For what will it profit a man if he gains the whole world and forfeits his soul?"

Luke 19:10: "For the Son of Man came to seek and to save the lost."

John 5:24: "Whoever hears my word and believes him who sent me has eternal life. He does not come into judgment, but has passed from death to life."

Romans 3:23: "For all have sinned and fall short of the glory of God."

Romans 6:23: "For the wages of sin is death."

Romans 5:8: "But God shows his love for us in that while we were still sinners, Christ died for us."

John 3:16: "For God so loved the world, that he gave his only Son, that whoever believes in him should not perish but have eternal life."

Matthew 7:13-14: "Enter by the narrow gate. For the gate is wide and the way is easy that leads to destruction, and those who enter by it are many. For the gate is narrow and the way is hard that leads to life, and those who find it are few."

John 10:9: "I am the door. If anyone enters by me, he will be saved and will go in and out and find pasture."

Romans 10:9: "If you confess with your mouth that Jesus is Lord and believe in your heart that God raised him from the dead, you will be saved."

Luke 15:10: "Just so, I tell you, there is joy before the angels of God over one sinner who repents."

Appendix
Is Heaven Real?

Matthew 5:3: Blessed are the poor in spirit, for theirs is the kingdom of heaven.

Matthew 5:12: Rejoice and be glad, for your reward is great in heaven, for so they persecuted the prophets who were before you.

John 14:2–3: In my Father's house are many rooms. If it were not so, would I have told you that I go to prepare a place for you? And if I go and prepare a place for you, I will come again and will take you to myself, that where I am you may be also.

Matthew 16:19: I will give you the keys of the kingdom of heaven, and whatever you bind on earth shall be bound in heaven, and whatever you loose on earth shall be loosed in heaven.

Matthew 18:3: Truly, I say to you, unless you turn and become like children, you will never enter the kingdom of heaven.

John 3:13: No one has ascended into heaven except he who descended from heaven, the Son of Man.

John 20:17: Jesus said to her, 'Do not cling to me, for I have not yet ascended to the Father; but go to my brothers and say to them, "I am ascending to my Father and your Father, to my God and your God."'

Acts 1:9: And when he had said these things, as they were looking on, he was lifted up, and a cloud took him out of their sight.

Mark 16:19: So then the Lord Jesus, after he had spoken to them, was taken up into heaven and sat down at the right hand of God.

Hebrews 9:24: For Christ has entered, not into holy places made with hands, which are copies of the true things, but into heaven itself, now to appear in the presence of God on our behalf.

Matthew 5:22: But I say to you that everyone who is angry with his brother will be liable to judgment; whoever insults his brother will be liable to the council; and whoever says, 'You fool!' will be liable to hell of fire.

Matthew 5:29–30: If your right eye causes you to sin, tear it out and throw it away. For it is better that you lose one of your members than that your whole body be thrown into hell. And if your right hand causes you to sin, cut it off and throw it away. For it is better that you lose one of your members than that your whole body go into hell.

Matthew 13:42: And throw them into the fiery furnace. In that place there will be weeping and gnashing of teeth.

Matthew 23:33: You serpents, you brood of vipers, how are you to escape being sentenced to hell?

Luke 16:23–24: And in Hades, being in torment, he lifted up his eyes and saw Abraham far off and Lazarus at his side. And he called out, 'Father Abraham, have mercy on me, and send Lazarus to dip the end of his finger in water and cool my tongue, for I am in anguish in this flame.'

Nicene Creeds
Original Nicene Creed

"We believe in one God, the Father Almighty, maker of all things visible and invisible.

And in one Lord Jesus Christ, the Son of God, begotten of the Father [the only-begotten; that is, from the essence of the Father], God from God, Light from Light, true God from true God, begotten not made, of one essence (*homoousios*) with the Father; through whom all things were made, both in heaven and on earth.

Who for us humans and for our salvation came down and was incarnate and became human, suffered and rose again on the third day, ascended into the heavens, and will come to judge the living and the dead.

And in the Holy Spirit.

But those who say, 'There was a time when he was not,' and 'Before he was begotten he was not,' and that 'He came into being from nothing,' or who affirm that the Son of God is of a different *hypostasis* or substance, or created, or is subject to alteration or change—these the Catholic and apostolic Church anathematizes."

Nicene Creed was expanded
by the Council of Constantinople

to expand te]he Holy Spirit section to affirm the Holy Spirit's divinity,
added more detail on Jesus' incarnation, crucifixion, and resurrection, and
include statements about the Church, baptism, and resurrection

"We believe in one God, the Father Almighty,
Maker of heaven and earth,
of all things visible and invisible.
And in one Lord Jesus Christ,
the only Son of God,
begotten from the Father before all ages,
God from God, Light from Light,
true God from true God,
begotten, not made;
of the same essence (*homoousios*) as the Father.
Through him all things were made.
For us and for our salvation
he came down from heaven;
he became incarnate by the Holy Spirit and the virgin Mary,
and was made human.
He was crucified for us under Pontius Pilate;
he suffered and was buried.
The third day he rose again, according to the Scriptures.
He ascended to heaven
and is seated at the right hand of the Father.
He will come again with glory
to judge the living and the dead.
His kingdom will never end.

And we believe in the Holy Spirit,
the Lord, the giver of life.
He proceeds from the Father
and with the Father and the Son is worshiped and glorified.
He spoke through the prophets.

We believe in one holy catholic and apostolic Church.
We affirm one baptism for the forgiveness of sins.
We look forward to the resurrection of the dead,
and to life in the world to come. Amen

Bibliography

Books and Scholarly Works:
- **Aland, K., & Aland, B.** *The Text of the New Testament: Its Transmission, Corruption, and Restoration* (E. F. Rhodes, Trans.). Eerdmans, 1995.
- **Albright, W. F.** *From the Stone Age to Christianity: Monotheism and the Historical Process.* Doubleday Anchor, 1957.
- **Allison, Dale C.** *Resurrecting Jesus: The Earliest Christian Tradition and Its Interpreters.* T&T Clark, 2005.
- **American Psychiatric Association.** *Diagnostic and Statistical Manual of Mental Disorders* (5th ed., text rev.). American Psychiatric Association Publishing, 2022.
- **American Bible Society.** *The Holy Bible: New International Version.* Zondervan, 2011.
- **Bagnall, R. S.** *Early Christian Books in Egypt.* Princeton University Press, 2009.
- **Barth, K.** *Church Dogmatics, Volume IV: The Doctrine of Reconciliation.* T&T Clark, 2004.
- **Bauckham, R.** *Jesus and the Eyewitnesses: The Gospels as Eyewitness Testimony.* Eerdmans, 2006.
- **Bauckham, Richard.** *Jesus and the God of Israel: God Crucified and Other Studies on the New Testament's Christology of Divine Identity.* Eerdmans, 2008.
- **Ben-Tor, A.** *The Archaeology of Ancient Israel.* Yale University Press, 2013.
- **Bietak, M.** *Avaris: The Capital of the Hyksos.* Oxford University Press, 2015.
- **Blayney, P. W. M.** *The First Folio of Shakespeare.* Folger Shakespeare Library, 1997.

- **Blomberg, C.** *The Historical Reliability of the Gospels.* IVP Academic, 1987.
- **Bryant Wood, J.** *Did the Israelites Conquer Jericho?* Biblical Archaeology Review, 16(2), 44-58, 1990.
- **Craig, W. L.** *The Son Rises: Historical Evidence for the Resurrection of Jesus.* Wipf & Stock, 2001.
- **Craig, W. L.** *Reasonable Faith: Christian Truth and Apologetics.* Crossway, 2008.
- **Collins, J. J.** *Beyond the Qumran Community: The Sectarian Movement of the Dead Sea Scrolls.* Eerdmans, 2010.
- **Crossan, J. D.** *The Historical Jesus: The Life of a Mediterranean Jewish Peasant.* Harper San Francisco, 1991.
- **Dunn, James D. G.** *Did the First Christians Worship Jesus? The New Testament Evidence.* Westminster John Knox Press, 2010.
- **Ehrman, Bart D.** *Did Jesus Exist? The Historical Argument for Jesus of Nazareth.* Harper One, 2012.
- **Ehrman, Bart D.** *Jesus: Apocalyptic Prophet of the New Millennium.* Oxford University Press, 1999.
- **Ehrman, Bart D.** *Jesus, Interrupted: Revealing the Hidden Contradictions in the Bible (And Why We Don't Know About Them).* Harper One, 2009.
- **Ehrman, Bart D.** *How Jesus Became God: The Exaltation of a Jewish Preacher from Galilee.* Harper One, 2014.
- **Evans, C. A.** *Jesus and His Contemporaries: Comparative Studies.* Brill, 2001.
- **Evans, C. A.** *Jesus and His World: The Archaeological Evidence.* Westminster John Knox Press, 2012.
- **Habermas, G. R., & Licona, M. R.** *The Case for the Resurrection of Jesus.* Kregel Publications, 2004.
- **Hurtado, Larry.** *Lord Jesus Christ: Devotion to Jesus in Earliest Christianity.* Eerdmans, 2003.

- **Josephus, Flavius.** *Antiquities of the Jews.* Translated by William Whiston, Hendrickson Publishers, 1987.
- **Lewis, C. S.** *Mere Christianity.* Harper One, 1952.
- **Licona, Michael R.** *The Resurrection of Jesus: A New Historiographical Approach.* InterVarsity Press, 2010.
- **McDowell, J.** *The New Evidence That Demands a Verdict.* Thomas Nelson, 1999.
- **McDowell, J.** *Evidence That Demands a Verdict: Life-Changing Truth for a Skeptical World.* Thomas Nelson, 2019.
- **Meier, J. P.** *A Marginal Jew: Rethinking the Historical Jesus, Vol. 1.* Yale University Press, 1991.
- **N.T. Wright.** *Jesus and the Victory of God.* Fortress Press, 1996.
- **N.T. Wright.** *The Resurrection of the Son of God.* Fortress Press, 2003.
- **Pliny the Younger.** *Letters of Pliny.* Translated by B. Radice, Penguin Classics, 1963.
- **Quran.** Various translations of Islamic texts.
- **Sanders, E. P.** *The Historical Figure of Jesus.* Penguin Books, 1993.
- **Sanders, E. P.** *Jesus and Judaism.* Fortress Press, 1985.
- **Schumaker, J. F.** *The Corruption of Reality: A Unified Theory of Religion, Hypnosis, and Psychopathology.* Prometheus Books, 1995.
- **Spitzer, R. J.** *New Proofs for the Existence of God: Contributions of Contemporary Physics and Philosophy.* Eerdmans, 2010.
- **Strobel, Lee.** *The Case for Christ: A Journalist's Personal Investigation of the Evidence for Jesus.* Zondervan, 1998.
- **Suetonius, Gaius.** *The Twelve Caesars.* Translated by Robert Graves, Penguin Classics, 2007.

- **Tacitus, Cornelius.** *Annals.* Translated by Alfred John Church & William Jackson Brodribb, Modern Library, 2003.
- **The Holy Bible.** *English Standard Version,* Crossway, 2022.
- **Witherington, Ben.** *The Jesus Quest: The Third Search for the Jew of Nazareth.* IVP Academic, 1997.

www.ingramcontent.com/pod-product-compliance
Lightning Source LLC
Chambersburg PA
CBHW031514040426
42445CB00009B/220